Behavior of cryogenic liquids under compensated gravity and non-isothermal boundary conditions

BEHAVIOR OF CRYOGENIC LIQUIDS UNDER COMPENSATED GRAVITY
AND NON-ISOTHERMAL BOUNDARY CONDITIONS

Vom Fachbereich Produktionstechnik

der

UNIVERSITÄT BREMEN

zur Erlangung des Grades
Doktor-Ingenieur
genehmigte

Dissertation

von

Dipl.-Phys. Nikolai Kulev

Gutachter:

Prof. Dr.-Ing. Michael Dreyer

Prof. Dr. Eberhard Bänsch
(Universität Erlangen)

Tag der mündlichen Prüfung: 13.06.2019

Bibliografische Information der Deutschen Nationalbibliothek
Die Deutsche Nationalbibliothek verzeichnet diese Publikation in der
Deutschen Nationalbibliografie; detaillierte bibliographische Daten
sind im Internet über http://dnb.d-nb.de abrufbar.
1. Aufl. - Göttingen: Cuvillier, 2020
   Zugl.: Bremen, Univ., Diss., 2019

   ISBN 978-3-7369-7152-3
   eISBN 978-3-7369-6152-4

ABSTRACT

This study deals with the investigation regarding the behavior of liquid cryogenic propellants in the rocket tanks under weightlessness with superheated tank walls. The study contributes in this way to the application-oriented basic research by describing mathematically and quantifying experimentally the influence of the evaporation on the reorientation of cryogenic liquids.

The motivation of this work is the improvement of the understanding of the complex fluid-dynamical and thermo-dynamical interfacial phenomena, occurring during the coasting phase of the cryogenic propellants, which influence the dynamics of the liquid and the vapor phase. The objective is the determination of the capillary-driven motion and position of the propellant taking into account the wall superheat. Drop tower experiments were performed and analyzed as well as scaling approaches were developed within the framework of this study. The focus was to capture the influence of evaporation from the interface and the contact line mathematically and in this way to extend the mathematical description of the reorientation. Liquid argon, methane and hydrogen were used during the performed experiments and varying wall temperature gradients above the liquid-vapor interface were applied.

The main assumption of this study is based on the mathematical principle that the solution of a partial differential equation depends on its boundary condition. In my case these are the boundary conditions at the interface and the contact line. These boundary conditions express the relation of the velocity to the mass flux due to evaporation. The other equations from the equation system describing the flow are influenced therewith. The resulting mathematical implication is that the solutions of the governing equations become functions of the parameters of these two boundary conditions. The empirical implication is a dependence of the interface motion on these parameters or on the dimensionless numbers arising from the boundary conditions.

As main results it can be summarized that an equation for the contact-line boundary condition was implemented in the mathematical description of the liquid motion which equation takes into account the evaporation and the wall superheat at the contact line, respectively. Both the contact-line velocity and the contact-line angle become functions of the evaporation at the contact line and accordingly of the related dimensionless numbers. The implementation of the contact-line boundary condition completes the equation system describing the flow macroscopically. A joint scaling approach of all field equations and boundary conditions was developed. The influence of evaporation on the capillary-driven motion was expressed mathematically through the scaling approach. This influence on the motion of the whole interface was experimentally quantified as well.

VI

The characteristics of the interface motion were presented and discussed, based on the above mentioned two boundary conditions, as dependencies of the dimensionless numbers related to the interfacial evaporation. This study describes the performed drop tower experiments and the scaling approaches required for their understanding. The conclusions resulting from these considerations are presented as well.

ZUSAMMENFASSUNG

Die vorliegende Arbeit beschäftigt sich mit den Untersuchungen zum Verhalten von flüssigen kryogenen Treibstoffen in Raketentanks unter Schwerelosigkeit und mit überhitzen Tankwänden. Somit leistet die Arbeit einen Beitrag zu den anwendungsbezogenen Grundlagenforschung, in dem der Einfluss der Verdampfung auf die Reorientiering kryogener Flüssigkeiten mathematisch erfasst und experimentell quantifiziert wird.

Die Motivation dieser Arbeit ist ein verbessertes Verständnis der komplexen fluiddynamischen und thermo-dynamischen Grenzflächenphänomene zu erarbeiten, die während der Coasting-Phase bei den kryogenen Treibstoffen auftreten, welche die Dynamik der Flüssigkeit und des Dampfes beeinflussen. Hintergrund ist es, die kapillar-getriebene Bewegung und die Position des Treibstoffs unter der Berücksichtigung der Wandüberhitzung zu bestimmen. Im Rahmen dieser Arbeit wurden Fallturmexperimente durchgeführt und analysiert sowie entsprechende Skalierungsansätze entwickelt. Der Fokus lag dabei, den Einfluss der Verdampfung von der Grenzfläche und von der Kontaktlinie auf die Bewegung der Grenzfläche mathematisch zu erfassen und somit die mathematische Beschreibung der Reorientierung zu erweitern. Für die im Rahmen dieser Arbeit durchgeführten Experimente wurden flüssiges Argon, Methan, Neon und Wasserstoff verwendet, wobei verschiedene Wandtemperaturgradienten oberhalb der Grenzfläche eingesetzt wurden.

Die Grundannahme dieser Arbeit beruht auf dem mathematischen Prinzip, dass die Lösung einer partiellen Differentialgleichung von derer Randbedingung abhängt. In unserem Fall sind das die Randbedingungen an der Grenzfläche und an der Kontaktlinie. Diese Randbedingungen stellen die Relation der Geschwindigkeit zu dem Massenstrom aufgrund der Verdampfung her. Dadurch werden andere Gleichungen von dem Gleichungssystem, welches die Strömung beschreibt, beeinflusst. Die mathematische Folgerung davon ist es, dass die Lösung der Grundgleichungen zur Funktion der Parameter dieser Randbedingungen wird. Als empirische Folgerung ergibt sich eine Abhängigkeit der Charakteristika der Grenzflächenbewegung von diesen Parametern oder von den dimensionslosen Kennzahlen, welche aus den Randbingungen enstehen.

Als wichtige Ergebnisse kann zusammengefasst werden, dass eine Gleichung für die Kontaktlinie-Randbedingung in der mathematischen Beschreibung der Flüssigkeitsbewegung implementiert wurde, welche die Verdampfung bzw. die Wandüberhitzung an der Randlinie explizit berücksichtigt. Dadurch werden die Geschwindigkeit der Kontaktlinie und der Randwinkel zur Funktionen der Verdampfung an der Kontaktlinie bzw. der damit verbundenen dimensionslosen Zahlen. Durch die Implemetierung der Kontaktlinie-Randbedingung wird einerseits das Gleichungssystem, welches die Strömung makroskopisch beschreibt, vervollständigt. Ein einheitlicher Skalierungsansatz für alle Feldgeichungen und Randbedingungen wurde entwickelt. Dadurch wurde der Einfluss der Verdampfung auf die kapillar-getriebene Bewegung mathematisch erfasst. Dieser Einfluss auf die Bewegung der gesamten Grenzfläche wurde auch experimentell quantifiziert.

Anhand der beiden oben genannten Randbedingungen wurden die Charakteristika der Grenzflächenbewegung in Abhängigkeit von den dimesionslosen Kennzahlen, welche mit der Verdampfung verbunden sind, vorgestellt und diskutiert. Die vorliegende Arbeit beschreibt die durchgeführten Fallturmexperimente und die Skalierungsansätze zu deren Verständniss. Die daraus gezogenen Schlussfolgerungen werden auch vorgestellt.

ACKNOWLEDGEMENTS

The study presented here resulted from my work as a researcher in the Fluid Mechanics and Multiphase Flow Group at the Center of Applied Space Technology (ZARM) at the University of Bremen. Therefore I would like, first of all, to express my great appreciation and gratitude to the head of this group Prof. Dr.-Ing. Michael Dreyer for all his scientific guidance, advice, supervision and patience. Without his support this work would not exist.

Furthermore, I want to thank Prof. Dr. Eberhard Bänsch from the University of Erlangen that he agreed to be the second supervisor of this thesis and for his participation in the examination process. I am thankful for his helpful advices during this process as well as for the fruitful collaboration for our joint paper.

I would also like to thank Prof. Dr.-Ing. Udo Fritsching and Dr.-Ing. Norbert Riefler for taking from their valuable time to be a part if the examination committee.

My special thanks are extended to my colleagues Peter Prengel and Frank Cecior for the design, integration and operation of the experimental setup. Without their commitment and technical mastery the preparation as well as the execution of the drop tower experiments would not have been possible.

Many thanks are due to my colleagues from the Fluid Mechanics and Multiphase Flow Group Dr.-Ing. Tim Arndt, Helmuth Behrens, Dr.-Ing. Yvonne Chen, Holger Faust, Eckart Fuhrmann, Dr.-Ing. Aleksander Grah, Dr.-Ing. Jörg Klatte, Dr.-Ing. Carina Ludwig and Dr.-Ing. Ming Zhang for the helpful discussions during my employment at ZARM.

I would like to thank Dr. Steffen Basting from the Technical University of Dortmund for his contributions to our joint paper, to my predecessor Malte Stief now at DLR Bremen for his contributions to development of the experimental setup and to my former student Florian Wolff for his contributions to the data evaluation, too.

The funding of the projects, from which my research activity was a part, by the German Federal Ministry of Economics and Technolodgy (BMWi) through the German Aerospace Center (DLR) under grant numbers 50 RL 0921 and 50 RL 1320 is gratefully acknowledged.

Last but not least I want to thank especially my precious wife Vessela Kulev. Without all her love, support, motivation, understanding and patience with me this thesis could not have been accomplished in the following manner.

Contents

List of Figures

List of Tables

Nomenclature

α	accommodation coefficient or ray propagation angle in vacuum regarding the center point
α^*	dimensionless difference between the final steady-state deflections
α_w	ray propagation angle in vacuum regarding the contact line
β	ray propagation angle in liquid regarding the center point
β^*	dimensionless curvature radius
β_w	ray propagation angle in liquid regarding the contact line
Δ	difference or jump of a quantity
δ_T	thickness of the thermal interfacial boundary layer, m
ε	aspect ratio
ε_{mn}	roots of the BESSEL function of first order
Γ	dimensional hysteresis parameter, m^{-1}
γ	ray propagation angle in the cylinder wall regarding the center point
Γ^*	dimensionless hysteresis parameter
γ_d	dynamic contact angle
Γ_I	surface area of the interface, m^2
γ_s	static contact angle
γ_w	ray propagation angle in the cylinder wall regarding the contact line
γ_{da}	advancing dynamic contact angle
γ_{dr}	receding dynamic contact angle
γ_{sa}	advancing static contact angle
γ_{sr}	receding static contact angle
γ_{ss}	steady-state contact angle
$\hat{\alpha}$	dimensionless coefficient, Eq.(2.36)
$\hat{\zeta}$	dimensionless damping ratio, Eq.(2.36) and Eq.3.80
κ	constant in the contact line boundary condition, $\mathrm{m\,s}^{-1}$
κ^0	equilibrium frequency of the molecular displacements, s^{-1}
Λ	mean logarithmic decrement
λ	thermal conductivity, $\mathrm{W\,m}^{-1}\,\mathrm{K}^{-1}$
λ_0	average distance of the molecular displacements, m
μ	liquid dynamic viscosity, Pa s
∇P	gradient of the pressure field, $\mathrm{hPa\,m}^{-1}$

∇T gradient of the temperature field, $\mathrm{K\,m^{-1}}$

$\nabla \cdot \mathbf{u}$ divergence of the velocity field, $\mathrm{s^{-1}}$

$\nabla \mathbf{u}$ gradient of the velocity field, $\mathrm{s^{-1}}$

$\nabla^2 T$ Laplace operator of the temperature field, $\mathrm{K\,m^{-2}}$

$\nabla^2 \mathbf{u}$ Laplace operator of the velocity field, $\mathrm{s^{-1}m^{-1}}$

ν liquid kinematic viscosity, $\mathrm{m^2\,s^{-1}}$

Ω_d dimensionless angular damped frequency

ω_d dimensional angular damped frequency, $\mathrm{s^{-1}}$

Ω_n dimensionless angular natural frequency

ω_n dimensional angular natural frequency in of the equivalent mechanical model of the reorientation through a step response, Eq.(3.79), $\mathrm{s^{-1}}$

ω_{01} dimensional angular natural frequency of the $m = 0$-th lateral and the $n = 1$-th axial sloshing mode corresponding to the reorientation, $\mathrm{s^{-1}}$

ω_{mn} dimensional angular natural frequency of the m-th lateral and the n-th axial sloshing mode, $\mathrm{s^{-1}}$

ρ liquid density, $\mathrm{kg\,m^{-3}}$

ρ_0 mean liquid density, $\mathrm{kg\,m^{-3}}$

ρ_v vapor density, $\mathrm{kg\,m^{-3}}$

$\rho_{v,\infty}$ vapor density far away from the interface, $\mathrm{kg\,m^{-3}}$

$\rho_{v,int}$ vapor density close to the interface, $\mathrm{kg\,m^{-3}}$

σ surface tension, $\mathrm{N\,m^{-2}}$

σ_T temperature gradient of the surface tension, $\mathrm{N\,m^{-2}\,K^{-1}}$

σ_{sl} surface tension at solid-liquid interface, $\mathrm{N\,m^{-2}}$

σ_{sv} surface tension at solid-vapor interface, $\mathrm{N\,m^{-2}}$

τ time scale, s

τ_v viscous time scale, s

Θ characteristic temperature difference in liquid, K

Θ_v characteristic temperature difference in vapor, K

Θ_w characteristic temperature difference in wall, K

ς adiabatic exponent

δA surface area increase of the interface, $\mathrm{m^2}$

a constant force input, Eq.(3.77), $\mathrm{kg\,m\,s^{-2}}$

a dimensionless parameter, Eq.(2.36)

b dimensionless parameter, Eq.(2.36)

c viscous damping constant, Eq.(3.77), $\mathrm{kg\,s^{-1}}$

$\hat{c}$ dimensionless compression rate

$\hat{C}^*$ dimensionless competition parameter, Eq. (2.34)

c compression rate, $\mathrm{s^{-1}}$

C^* dimensionless competition parameter, Eq. (2.32)

c_p specific heat capacity at constant pressure, $\mathrm{J\,kg^{-1}\,K^{-1}}$

c_v specific heat capacity at constant volume, $\mathrm{J\,kg^{-1}\,K^{-1}}$

$\hat{D}$	damping ratio
$\mathbf{f}$	external body forces, $\mathrm{kg\,m\,s^{-2}}$
D	diameter, m
D_m	diffusion coefficient, $\mathrm{m^2\,s^{-1}}$
δG	change of the free energy, J
g	acceleration due to gravity on Earth, $\mathrm{m\,s^{-2}}$
Δh_{lv}	latent heat of evaporation, $\mathrm{J\,kg^{-1}}$
h	height function of the interface, m
h^*, h^+	dimensionless height of the interface
h_0	initial height of the interface (fill level), m
h_{av}	cross-sectional averaged height of the interface, m
$h_{ss,i}$	isothermal steady-state height of the interface, m
h_{ss}	steady-state height of the interface, m
j, J	interfacial mass flux and corresponding scale, $\mathrm{kg\,s^{-1}}$
J_0	evaporation parameter (diffusion coefficient), $\mathrm{m^2\,s^{-1}}$
j_{cl}, J_{cl}	interfacial mass flux at the contact line corresponding scale, $\mathrm{kg\,s^{-1}}$
K	non-equilibrium parameter, $\mathrm{kg^{-1}\,s\,K}$
k	mean interface curvature, $\mathrm{m^{-1}}$
k	spring constant of the equivalent mechanical model of the reorientation through a step response, Eq.(3.77), $\mathrm{kg\,s^{-2}}$
k_B	Boltzmann constant, $\mathrm{m^2\,kg\,s^{-2}\,K^{-1}}$
k_{01}	spring constant of the equivalent mechanical model of the axisymmetric $m = 0$-th lateral and the $n = 0$-th axial sloshing mode corresponding to the reorientation, Eq.(3.83), $\mathrm{kg\,s^{-2}}$
k_{mn}	spring constant of the equivalent mechanical model of the m-th lateral and the n-th axial sloshing mode, Eq.(3.82), $\mathrm{kg\,s^{-2}}$
k_{zi}	acceleration due to gravity, $\mathrm{m\,s^{-2}}$
L	characteristic length, m
m	mobility exponent
m	moving mass of the equivalent mechanical model of the reorientation through a step response, Eq.(3.77), kg
m_0	rigid (motionless) mass of the equivalent mechanical model of the reorientation step response, kg
M_m	molar mass, $\mathrm{kg\,mol^{-1}}$
m_{01}	mass of the equivalent mechanical model of the axisymmetric $m = 0$-th lateral and the $n = 1$-th axial sloshing mode corresponding to the reorientation, Eq.(3.83), kg
m_{mn}	sloshing mass of the equivalent mechanical model of the m-th lateral and the n-th axial sloshing mode, Eq.(3.82), kg
n	refractive index of the liquid

n_{vac}	refractive index of vacuum
n_w	refractive index of the cylinder wall
P	liquid pressure, hPa
P_A	ambient pressure, hPa
P_A	gas pressure, hPa
P_S	saturation pressure, hPa
P_v	vapor pressure, hPa
$P_{v,hyd}$	hydrodynamical vapor pressure, hPa
$\dot{q}$	specific heat flux, $\mathrm{W\,m^{-2}}$
R	cylinder radius, m
R_1, R_2	principal curvature radii, m
R_m	mean curvature radius, m
R_s	specific gas constant, $\mathrm{J\,kg^{-1}\,K^{-1}}$
R_u	universal gas constant, $\mathrm{J\,mol^{-1}\,K^{-1}}$
RES_{cl}	resolution of the experimental images related to the contact line, px/mm
RES_{cp}	resolution of the experimental images related to the center point, px/mm
$\mathbf{T}$	Cauchy stress tensor, Pa
T	liquid temperature, K
T_I	interface temperature, K
T_S	saturation temperature, K
T_v	vapor temperature, K
T_w	wall temperature, K
T_{ch}	characteristic liquid temperature, K
t_{cp}	time of the maximum deflection of the center point, s
t_{Lc}	capillary time scale, s
t_{pu}	unsteady pressure time scale, s
t_p	time of the peak deflection of the moving mass of the equivalent mechanical model, s
t_{s1}	time needed by the center point to pass its new steady-state position, s
t_s	settling time of the center point, s
t_T	time of buildup of the interfacial thermal boundary layer, s
$T_{v,ch}$	characteristic vapor temperature, K
$T_{w,ch}$	characteristic wall temperature, K
t_{wp}	time of the maximum deflection of the contact line point, s
$\mathbf{u}$	vector of liquid velocity, $\mathrm{m\,s^{-1}}$
$\mathbf{u}_v$	vector of vapor velocity, $\mathrm{m\,s^{-1}}$
U	characteristic liquid velocity, $\mathrm{m\,s^{-1}}$
$u\,|_{cl}$	liquid velocity at the contact line, $\mathrm{m\,s^{-1}}$
$u\,|_{cp}$	liquid velocity at the center point, $\mathrm{m\,s^{-1}}$
u_{av}	average axial velocity of the liquid, $\mathrm{m\,s^{-1}}$
u_{cl}	contact line velocity, $\mathrm{m\,s^{-1}}$
u_{cp}	center point velocity, $\mathrm{m\,s^{-1}}$

u_{ev}	evaporative velocity scale, $\mathrm{m\,s^{-1}}$
$u_{I,cl}$	normal interface velocity at the contact line, $\mathrm{m\,s^{-1}}$
u_I	normal interface velocity, $\mathrm{m\,s^{-1}}$
u_{puLc}	capillary velocity scale with a length scale L_c, $\mathrm{m\,s^{-1}}$
u_{puL}	capillary velocity scale with a length scale L, $\mathrm{m\,s^{-1}}$
u_{puR}	capillary velocity scale with a length scale R, $\mathrm{m\,s^{-1}}$
$u_{r,m}$	non-dimensionalized average rise velocity of the contact line
$u_s(t)$	unit step function, Eq.(3.77)
u_v	vapor velocity, $\mathrm{m\,s^{-1}}$
Z	deflection of the moving mass of the equivalent mechanical model of the reorientation from its initial position, Fig. (3.2) and Eq. (3.77), m
Z_p	maximal deflection of the moving mass of the equivalent mechanical model of the reorientation from its initial position, Eq.(3.90), m
Z_{ss}	steady-state deflection of the moving mass of the equivalent mechanical model of the reorientation from its initial position, Eq. (3.85), m
Z_c	deflection of the interface center point, m
Z_w	deflection of the contact line, m
Z_{cp}	maximal deflection of the center point, m
Z_{css}	final steady-state deflection of the center point, m
Z_{w0}	initial deflection of the contact line point, m
Z_{wp}	maximal deflection of the contact line point, m
Z_{wss}	final steady-state deflection of the contact line point, m
Bo	Bond number
Ca	Capillary number
$\mathbf{E_{cl}}$	Evaporation number at the contact line
$\mathbf{Ev_v}$	Evaporation heat flux number in the vapor
Ev	Evaporation heat flux number
E	Evaporation number
Ja	Jakob number
$\mathbf{Oh_v}$	Ohnesorge number in the vapor
Oh	Ohnesorge number
$\mathbf{Pe_w}$	Peclet number the wall
$\mathbf{Pr_v}$	Prandtl number in the vapor
Pr	Prandtl number
$\mathbf{Ptd_v}$	Thermodynamic Pressure number in the vapor
$\mathbf{R_\rho}$	Density ratio number
$\mathbf{Re_M}$	Thermocapillary Reynolds number
We	Weber number

1 Introduction

Understanding the behavior of cryogenic liquids under reduced gravity and thermal influences from the environment is of great importance in the management of cryogenic propellants of the next-generation launchers. After the end of thrust of the engines the propellant is driven by the now dominating capillary forces from the steady-state position under normal gravity to a new steady-state position under microgravity, Fig. 1.1. Until the new steady-state position is reached, the interface exhibits decaying damped oscillations typical of underdamped systems and the whole process is known as interface (free surface) reorientation. During the reorientation the propellant moves along the warmer wall. Knowing the position of the propellant interface as well as pressure and temperature evolutions is crucial in handling the propellant during this ballistic phase.

The non-isothermal reorientation can be summarized in one sentence as following: step response of the liquid-vapor interface through capillary-driven flow with interfacial phase change. The general topic of the present work is the influence of the mass and heat transport on the linear impulse transport during the reorientation of the interface in a partially filled container under compensated gravity. So in specific terms the above summary of the reorientation can be reformulated as the influence of the interfacial mass flux on the interface dynamics and steady-state under μg through the corresponding boundary conditions at the interface and the contact line.

During the reorientation a transition takes place of the interface from its 1g steady-state position under the action of gravity to its μg steady-state position under the action of capillarity and evaporation. Therefore I use two types of the mathematical description of the reorientation, Fig. (1.2). On the one side it is the fluid-dynamical description of capillary-driven flow with interfacial phase change, and with a contact line boundary condition explicitly accounting for the evaporation due to the wall superheat. On the other side there is the treatment of the phenomenon in the terms of the system dynamics - as a linear step response of 2nd order as in [58]. Thereby the focal point is the influence of the non-isothermal conditions on the fluid-dynamical and the system-dynamical quantities and parameters describing the reorientation.

After the step reduction of gravity the value of the contact angle is no longer equal to the value of the static contact angle γ_s but to the value of the dynamic contact angle γ_d. On the other side a reduction of the pressure occurs at the contact line due to the disappearance of the hydrostatic forces. So a capillary-driven motion

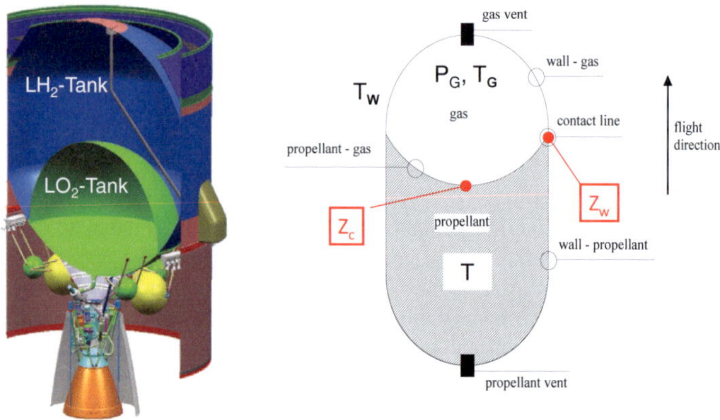

Fig. 1.1. Rendering of the Ariane 5 upper stage and a generic sketch of its tank. The tank wall, the propellant and the gas in the ullage have temperatures T_w, T and T_g, respectively, and the gas pressure is P_g. The points of special interest from the interface between the propellant ant the gas are contact line with a coordinate Z_w and the center point with a coordinate Z_c.

arises at the contact line due to both these reasons: the contact line begins moving towards its steady-state position corresponding to γ_s simultaneously with a flow from the liquid bulk to the contact line region due the pressure difference. Regarding the contact line boundary condition, containing the velocity of the contact line u_{cl} in Fig. (1.2), the difference between the dynamic and static contact angles $\gamma_d - \gamma_s$ can be considered as a variable describing the capillary-driven flow, [17] and [8].

In this way the contact line is moved upwards the wall to regions where a significant difference between the wall and interface temperatures exists. At this point the action of the evaporation superimposes the capillary-driven motion. The mass flux at the contact line j_{cl} is the variable describing the action of evaporation on its motion [2]. So the contact line velocity u_{cl} becomes influenced by the the evaporative term j_{cl}, too, besides the non-evaporative term $\gamma_d - \gamma_s$. In this manner the motion and the steady-state of the contact line become functions of the mass flux j_{cl}, too, when interfacial evaporation is present. The new steady-state is characterized by the contact angle γ_{ss} which is determined through the contact line boundary condition. This condition was used by [2] for the investigation of the influence of the evaporation from the interface and the contact line on the droplet spreading.

The motion of the center point is expressed through the interface boundary condition at the center point containing the velocity of the center point u_{cp} in Fig. (1.2). This velocity would be influenced by both the non-isothermal term with the mass flux at the center point j_{cp} and the isothermal term with the liquid velocity there $u\,|_{cp}$. This boundary condition is the equivalent of the mass balance boundary condition at the interface given by Eq. (2.27) and used by [64] for the investigation of the influence of the interfacial phase change on the capillary rise.

Due to the departure of the interface curvature from its steady-state curvature under isothermal conditions during the reorientation I assume that a capillary-driven flow with the characteristic velocity u_{puR} in Fig. (1.2) arises [81], [19], [50] which is the cause for the motions of the liquid and the interface.

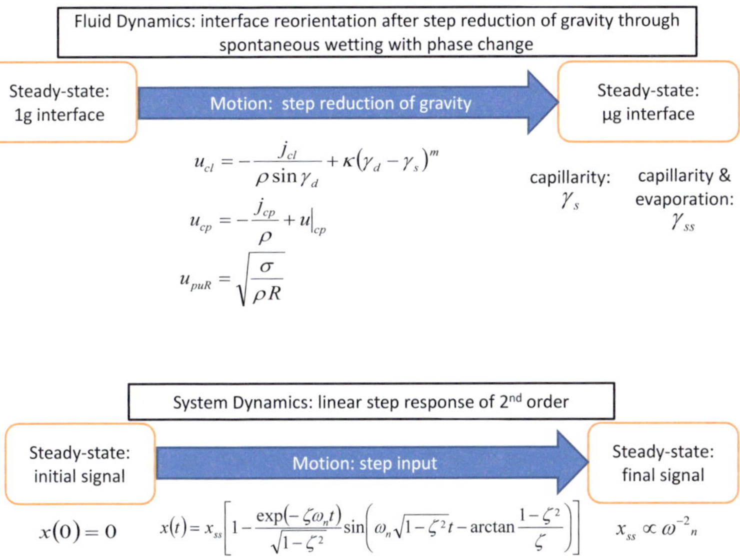

$$u_{cl} = -\frac{j_{cl}}{\rho \sin \gamma_d} + \kappa (\gamma_d - \gamma_s)^m$$

$$u_{cp} = -\frac{j_{cp}}{\rho} + u\Big|_{cp}$$

$$u_{puR} = \sqrt{\frac{\sigma}{\rho R}}$$

$$x(0) = 0 \qquad x(t) = x_{ss}\left[1 - \frac{\exp(-\zeta\omega_n t)}{\sqrt{1-\zeta^2}}\sin\left(\omega_n\sqrt{1-\zeta^2}\,t - \arctan\frac{1-\zeta^2}{\zeta}\right)\right] \qquad x_{ss} \propto \omega^{-2}{}_n$$

Fig. 1.2. Basic features of the mathematical descriptions of the reorientation. Fluid dynamics: a motion due to capillary-driven flow with characteristic velocity u_{puR} with the interfacial phase change (mass fluxes j_{cl} and j_{cp}), leading to a change of both the interface dynamics (velocities u_{cl} and u_{cp}) and the µg steady-state (contact angle γ_{ss} instead of γ_s). System dynamics: the square of the frequency ω_n of the damped oscillations, until the final steady-state is reached, is inversely proportional to the steady-state deflection x_{ss}.

Therefore the presence of non-zero mass flux at the interface would turn the non-isothermal reorientation into a capillary-driven flow phenomenon with interfacial phase change. The influence of the interfacial phase change is manifested by variation of the interfacial mass flux in the corresponding boundary conditions at the interface and the contact line due to the variation of the axial wall temperature gradients. This assumption is based also on the examples of the capillary-driven flow with interfacial phase change such as capillary rise and droplet spreading, which I summarize in Section. (2.5). The characteristic feature of these examples is the existence of the isothermal values of the quantitative characteristics of the dynamics and the steady-state of the interface. In the presence of interfacial phase change these values vary with the variation of the interfacial mass flux.

The above mentioned examples in the literature also show that action of evaporation is opposite to the action of capillarity [2], [64]. This means that evaporation either increases or decreases the above mentioned isothermal values, achieved solely under the action of capillarity. So the isothermal values of the reorientation characteristics based on the investigation on [50] should vary accordingly, too. Based on [2] and [64] the interface dynamics be interpreted as the mentioned competition between the capillarity and the evaporation, whereas the interface steady state can be regarded as their balance.

From the viewpoint of the system dynamics the reorientation can be approximated by a step response of a linear system of second order and the characteristics of the interface motion can be treated as the characteristics of this step response. The damped oscillations about the steady state deflection of such a system process exhibit a natural frequency ω_n which is mathematically related to this deflection x_{ss} as $\omega_n^2 \propto x_{ss}^{-1}$ [58], shown in Fig. (1.2). A variation of the wall superheat would result in a variation of the steady-state contact angle which in turn would lead to a variation of the steady-state deflection. Given the aforementioned relation between frequency and deflection a corresponding variation of the frequency would be expected too. So both the transient quantity (frequency) and the steady-state quantity (deflection) become functions of the wall superheat.

Up to the present moment the investigation of the liquid reorientation has been largely restricted to storable liquids in partly filled cylinders primarily only under isothermal conditions expressed by an adiabatic non-heated wall. The main results were achieved by [50] through experiments and by [29] through numerical simulations. The present study aims at the extension of the aforementioned results to the range of cryogenic liquids under non-isothermal conditions in the presence of a superheated wall. The influence of the interfacial mass flux on the reorientation characteristics is expected to manifest itself through the boundary conditions at the interface and the contact line, as already mentioned. The results by [50] and [29] show the influence of the contact line on the overall flow pattern in the liquid and on motion of the whole interface. Accordingly the variation of the contact line condition through the superheated wall should impact both the liquid flow pattern and the interface motion.

The present work has two major goals. The first one comprises the further development of the mathematical description of the reorientation used by [50] and [29] for the isothermal one-phase case to the non-isothermal two-phase case with pure liquids accounting for the wall superheat and the identification of the relevant dimensionless numbers describing the influence of the interfacial phase change on the dynamics through the boundary conditions at the interface and the contact line. The second one aims at the experimental quantification of the influence of the variation of the mass flux at the interface and the contact line on the reorientation characteristics and the identification of the dependencies of these characteristics on the aforementioned dimensionless numbers.

2 State of the art and research goals

As already stated, the object of the present research is the reorientation of the liquid-vapor interface of the single-species, two-phase cryogenic liquids systems under non-isothermal wall boundary conditions and compensated gravity. The motion of an interface with phase change is investigated by its transition from one steady-state configuration under normal gravity to another steady-state position under microgravity through the capillary-driven process of spontaneous wetting.

The present work is the extension of the investigations by [50] and [29] of the isothermal reorientation. However, there is a clear connection of the present investigation to the physical mechanisms, mathematical formalisms and the results of other investigations as well. Accordingly, the task of present chapter is to present this connection, too, which can be summarized as: the influence of the boundary conditions at the interface and at the contact line on the interface dynamics in cases of a capillary-driven interfacial flow [81], both with and without phase change. In other words the variation in these boundary conditions would lead to a variation of the interface dynamics. This is apparent since the interface velocity appears explicitly in these boundary conditions. Accordingly the conditions can be viewed as a functional dependence of the interface velocity on the other dynamic quantities and constant parameters. So the above mentioned influence is expressed by these quantities and parameters or by the respective dimensionless numbers from the boundary conditions. Therefore the physical quantities and relations needed for both the definition of the boundary conditions at the interface and the contact line as well as the field equations presented in the next chapter are introduced here. This chapter presents also the research goals of the present work.

Physical mechanisms related to the liquid-vapor interface through the corresponding boundary conditions essentially influence the flow and the interface motion during the interface reorientation [81], [50], [29]. The knowledge of the action of these mechanisms and especially of their interaction is the foundation for the understanding of the reorientation process under non-isothermal wall boundary conditions. The two mechanisms which are considered here are capillarity and evaporation. The interaction between these mechanisms in terms of superposition and competition is assumed by me to be the basic underlying mechanism behind the non-isothermal reorientation based on [2] and [13]. The interaction between capillarity and evaporation can be quantitatively described by certain relations presented in this chapter.

Accordingly this chapter has the following structure. The steady state of the interface under capillary conditions is presented in Section (2.1). The definitions of interfaces, surface tension, static capillary interfaces and static contact angles are introduced there.

The contact line boundary conditions under isothermal conditions are treated in the next Section (2.2). For that purpose the dynamic contact angle and the dynamic contact angle relations are introduced there. The influence of the contact line boundary condition on the interface dynamics by the isothermal droplet spreading is given there. This influence by the isothermal reorientation and the isothermal damped interface oscillations is discussed in Section (2.6).

The phase change at the interface and the contact line is treated in Section (2.3). Relations of the interfacial mass flux due to the phase change (evaporation or condensation) are presented which are used for the formulation of the boundary conditions at the contact line in Section (2.4) and at the interface in Section (3.3).

The contact line boundary conditions with interfacial phase change are presented in Section (2.4). The steady-state value of the dynamic contact angle can be computed by one of these relations chosen by me as the boundary condition for the field equations in Section (3.3).

The capillary-driven flow with interfacial phase change is considered in Section 2.5. The influence of the phase change (mass flux or corresponding temperature difference) on the interface dynamics through the boundary conditions at the interface and the contact line is presented by the capillary rise and droplet spreading. The competition between capillarity and evaporation and its quantification are presented alongside with this.

Previous results concerning the reorientation of the interface from normal gravity to compensated gravity conditions and the damped interface oscillations under compensated gravity are treated in Section (2.6). As mentioned above the influence of the boundary conditions at the interface and the contact line on the interface dynamics are discussed in this section.

Finally, a synopsis and the goals of the present research are formulated in Section (2.7).

2.1 Steady-state interfaces under isothermal conditions.

As already mentioned the steady-state interface under capillary conditions are characterized by the static equilibrium of forces acting on the interface by the absence of any flow. Accordingly this interface can be termed as static, too. Under these conditions the contact line formed by the intersection of the interface with the solid boundary is also static. The angle of this intersection has a constant value which is termed static contact angle. The relations describing mathematically this physical situation follow.

2.1.1 Steady-state interfaces.

According to [10] an interface between the fluid phases is the boundary between the separate phases which under microscopic observation exhibits a finite thickness and is designated as interface boundary layer. The interface is characterized by a drastic change of a physical property (thermo-physical, optical, etc.) within the interface boundary layer from the value in the first phase to the value of the second phase. The thickness of the interface boundary layer between liquid and vapor at saturation is around 1 nm for normal liquids. For the purposes of my macroscopic investigation this thickness is much smaller than any relevant length scale so the interface can be regarded as a mathematical surface of an infinitesimal thickness, across which the physical properties of the phases exhibit a discontinuity. Physical properties are assigned to the interface which can be interpreted as mean values within the thickness of the interface boundary layer. Here I must point out that the boundary layer representation of the interface itself is entirely different from the microscopical interfacial boundary layers, both velocity and temperature, characterized by much greater thicknesses, with length scales of around 1 mm and more.

The fundamental physical property which characterizes the interface macroscopically and relates it to capillarity is surface tension. The thermodynamical definition of surface tension σ reads

$$\sigma = \frac{\delta G}{\delta A} \tag{2.1}$$

where δG is the change of the free energy of the system required for the increase δA of the surface area of the interface [57].

When there is no flow under the action of an volume force (gravity) the interface shape is governed by the Young-Gauss-Laplace equation relating the pressure difference across the static interface and its mean curvature [1]

$$P - P_A = \sigma \left(\frac{1}{R_1} + \frac{1}{R_2} \right) \tag{2.2}$$

where P is the pressure on the liquid side and P_A is the ambient pressure. The quantities R_1 and R_2 are the principal curvature radii of the interface. The equation expresses the balance of normal stress to the interface by the lack of a tangential stress.

The quantity $1/R_1 + 1/R_2$ gives the mean curvature of the interface with the corresponding radius $1/R_m$. The mean curvature radius is given [81], [50] for an arbitrary interface parametrized as $h = h(x,y)$ in Cartesian coordinates by

$$\frac{1}{R_m} = \pm \frac{(1+q^2)s - 2pqv + (1+p^2)t}{(1+p^2+q^2)^{3/2}} \tag{2.3}$$

where $p = \partial h/\partial x$, $q = \partial h/\partial y$, $s = \partial p/\partial x$, $t = \partial q/\partial y$ and $v = \partial p/\partial y = \partial q/\partial x$.

Apparently the Young-Gauss-Laplace equation is so a nonlinear, partial differential equation containing second order spatial derivatives of the parametrization function of the interface. Accordingly one boundary condition is required at the edge of a interface to have a well-posed problem. Usually this boundary condition is given by the static contact angle, prescribed where the meniscus intersects a solid surface, or seldom by the position of the contact line of the intersection [81], [17].

2.1.2 Static contact lines and angles

A contact line is formed at the intersection of two immiscible fluids and a solid. The mutual interaction between the three materials in the immediate vicinity of a contact line can significantly affect the statics as well as the dynamics of an entire flow field [21]. Under ideal conditions, a liquid at rest intersects a solid at a unique angle. The static contact angle γ_s is the angle at which the meniscus intersects the solid as measured through the liquid. In the case of a pure liquid the Young equation is valid in a thermodynamic equilibrium [28] giving the value of γ_s:

$$\sigma_{sv} - \sigma_{sl} = \sigma \cos \gamma_s \qquad (2.4)$$

where σ_{sv} and σ_{sl} are the surface tensions of the solid-liquid and the solid-vapor interfaces, respectively. The above equation expresses the state of mechanical equilibrium due to the the balance forces at the contact line. A special case of Eq. (2.4) is the complete wetting liquids with $\gamma_s = 0°$ ($\sigma_{sv} - \sigma_{sl} = \sigma$), to which case the cryogenic liquids investigated here belong. In this case a liquid film of macroscopic thickness would be formed as an equilibrium interface on the solid [28]. Partially wetting liquids exhibit $0° < \gamma_s < 90°$ and non-wetting liquids - $90° < \gamma_s < 180°$. In both cases $\sigma_{sv} - \sigma_{sl} < \sigma$.

It is found that the contact line remains motionless not just for the contact angle value $\gamma = \gamma_s$ but for the interval $\gamma_{sr} \leq \gamma \leq \gamma_{sa}$ [21]. A motion of the contact line occurs only if the contact angle value lies outside of the above interval. So the biggest value and the smallest value of the contact angle by which a motionless contact line is present are denoted as γ_{sa}, advancing static contact angle, and γ_{sr}, receding static contact angle, respectively. If the contact angle takes values $\gamma > \gamma_{sa}$ an advancing contact line motion is initiated or wetting. By values of the contact angle $\gamma < \gamma_{sr}$ a receding contact line motion begins and so dewetting. The presence of the interval $[\gamma_{sr}, \gamma_{sa}]$ is known as contact-line hysteresis. It depends on the surface roughness of the solid, chemical contamination in the solid surface, solutes in the liquid [28].

2.2 Contact line boundary conditions under isothermal conditions.

The wetting process, the spreading of liquid over a smooth solid, represents a fundamental problem in fluid mechanics. It exemplifies the general problem of moving contact lines, which is encountered in such applications as the performance in Space of fuel tanks, mould filling and coating technology [24].

So the wetting behavior in the isothermal case is encoded with the constitutive equation relating the contact line velocity to the dynamic contact angle alongside with some material parameters, among which the static constant angle is always present. This relation has to be specified since at the contact line the chemistry of the surfaces has an effect [24]. On the other side the field equations describing the interfacial flow demand for their closure a boundary condition at the contact line given by such a relation [20], [81]. So the equation gives a contact line effect of the mathematical description of the investigated wetting phenomenon [24]. In this way both the mathematical form and the parameter values of the contact line boundary condition affect the interface dynamics.

Besides the above mentioned dimensional form some of the empirical or semi-empirical relations of the contact line boundary condition are given as functional dependence between the dynamic contact angle and the CAPILLARY number $\mathbf{Ca}$

$$\mathbf{Ca} = \frac{\mu U}{\sigma} \tag{2.5}$$

where μ and σ are the dynamic viscosity of the liquid and the surface tension. The velocity scale of the liquid flow is U.

2.2.1 Moving contact lines and dynamic contact angles

The problem of the moving contact line appears by moving or forming interfaces on solid surfaces, [70]. The motion of the contact line results from the lack of equilibrium at the contact point, described in Sec. 2.1. This lack of equilibrium is either caused by the application of external forces or is manifested by the unsteady migration of the liquid over a solid towards equilibrium, [5], [6], [42]. In the first process, referred to as *forced wetting*, the external (hydrodynamical or mechanical) forces lead to an increase of the the interfacial area beyond the conditions of static equilibrium. A typical example of the forced wetting is continuous deposition of a thin liquid layer onto a solid surface, known as coating. The second process is the *spontaneous spreading* of a liquid driven not by external forces but by the liquid/solid interactions resulting in the tendency of the liquid to relax towards an equilibrium through the reduction of its free energy by increasing of the wetted solid area. Examples of the spontaneous spreading are the imbibition of porous media and flotation. The interface reorientation is also an example of the spontaneous spreading. Both the constitute the *dynamic wetting* - the displacement of a fluid by a liquid from a solid surface [42], in contrast to static wetting characterize by the static contact angle [81].

Since the conditions during the dynamic wetting are non-equilibrium, the angle formed by the moving interface and the solid at the contact line position is no longer the static contact angle, presented in Sec. 2.1. The contact angle during the dynamic wetting is referred to as the *dynamic contact angle* γ_d. By an advancing contact line, observed by the initial wetting of the solid surface, one speaks of the advancing dynamic contact angle γ_{da}. A receding contact line, occurring by the de-wetting of the solid surface, is characterized by the receding dynamic contact angle γ_{dr}. The dynamic contact angle is a transient quantity essentially different from the static contact angle γ_s characterized by a single value, [5] [6].

The contact line velocity and the dynamic contact angle are the main quantities used for the quantification the dynamic wetting (in the isothermal case) [6] and a general monotonic relation between them is observed. This relation is regarded as the key boundary condition affecting the dynamic wetting, [6].

Some physical effects affect the process of the dynamic wetting and therefore influence the contact line dynamics: evaporation/condensation, thermocapillary and solutocapillary convection, solidification. As a result the contact line velocity and the dynamic contact angle become related to other quantities, characteristic of these effects. For the treatment of the present research object, the interface reorientation of single-species, two-phase cryogenic liquids systems with heat transfer from the wall, I concentrate myself on the case with evaporation/condensation this quantity is the temperature differences between the liquid and wall or the interfacial mass flux at the position of the contact angle.

The moving contact line demands the explicit definition of two relations serving as boundary conditions which are needed for the closure of the field equations in Section (3.3): the contact-line boundary condition or the dynamic contact angle relation and a slip boundary condition (the slip law).

The contact-line boundary condition is the relationship between the contact line velocity and the dynamic contact angle where some other parameters of the system such as the static contact angle or the wall surface roughness can have influence in the isothermal case. This isothermal case, without heat transfer from the wall, are presented in Sec.2.2.2. The respective relations for the non-isothermal case with interfacial phase change and heat transfer from the wall, are presented in Sec.2.4.1.

The slip boundary condition defines whether the flowing fluid adheres to the wall or slides at the wall. The former case leads to singularities of the wall shear stress and fluid velocity at the contact line while the latter case leads to finite values. So the slip law relieves the singularity. However, no data is known to me that slip law dep temperature - no relevant dimensionless numbers are expected from for the description of the experimental results. More importantly, my mathematical formalism is an extension of approaches which do not include explicitly the slip law and as mentioned above the dynamic contact angle relation is regarded as the key boundary condition affecting the dynamic wetting, [6]. Therefore I do not present here any slip law relations and I do not include the slip law within the field equation and the boundary conditions in Section (3.3).

So the aim of the present work is to take into account the evaporation process at the contact line macroscopically as done by [2]. Accordingly such a relation of the macroscopic dynamic contact angle is sought as a contact line boundary condition. The corresponding macroscopic scales are of at least several tens of microns ($L > 10\,\mu$m) in contrast to submicroscopic scales ($L \approx 3 \times 10^{-3} - 10^{-1}\mu$m) and molecular scales [42]. On these macroscopic scales, accessible to the naked eye or to optical microscopes, the contact line is identified where the liquid interface seems to intersect the solid surface a measurable dynamic contact angle γ_d.

2.2.2 Relations for the dynamic contact angle and the contact line motion.

Here I want to present primarily such dynamic contact angle relations in the isothermal case which have been which used as contact line boundary conditions for investigations of the the interface dynamics. The contact line boundary condition has a significant influence on the this dynamics by problems related to the spontaneous wetting/spreading:[83], [29], [55] for the isothermal reorientation, [40] for the interface oscillations, [33], [25], [2] for the droplet spreading. The investigations of the isothermal reorientation and interface oscillations are discussed in Section (2.6.1) whereas the droplet spreading is discussed here.

The seminal overview article on the treatment of the moving contact line was given by [21]. Based on various experimental data it was discussed by [21] how the contact line velocity depends on the dynamic contact angle. Following this discussion such a dependence was proposed by [25] considering an uni-directional contact line motion with a velocity u_{cl}

$$u_{cl} = \kappa \left(\gamma_{da} - \gamma_{sa} \right)^m \tag{2.6}$$

where $\kappa > 0$ is a empirical constant, $\gamma_{sa} \geq 0$ is the advancing static angle and $m \geq 1$ is another empirical constant, namely the mobility exponent. The form of Eq. (2.6) for $m = 1$ was proposed by [32] and for $m = 3$ is in agreement with the data on the contact line dynamics by [68], [36], [77] and [24]. Large values of κ means that the contact line is very mobile. By a decrease of κ the motion of the contact line is retarded so that very small values of κ would correspond to almost immobile contact line. Accordingly, the inverse value κ^{-1} is identified as measure of the contact line dissipation [25], [16].

Equation (2.6) was used by [25] for the investigation of the isothermal droplet spreading as a contact line boundary condition within the lubrication approximation of the Navier-Stokes equations. The value $m = 3$ gives an excellent agreement between the experimental and the theoretical results for the capillary-dominated spreading.

An investigation of the influence of the contact line boundary condition on the isothermal droplet spreading was performed by [33], too. For the purpose Eq. (2.6) was used with values of $m \in [1, 2, 3]$ and the lubrication approximation of the Navier-Stokes equations. It was found that the spreading rates strongly depend on the form of the contact line boundary condition, i.e on the value of m and the static contact angle γ_{sa}.

A generalization of Eq. (2.6) accounting for the receding contact line motion and the hysteresis was proposed by [2]

$$u_{cl} = \begin{cases} \kappa \left(\gamma_d - \gamma_{sa} \right)^m & , \ \gamma_d > \gamma_{sa} \\ 0 & , \ \gamma_{sr} \leq \gamma_d \leq \gamma_{sa} \\ \kappa \left(\gamma_d - \gamma_{sr} \right)^m & , \ \gamma_d < \gamma_{sr} \end{cases} \tag{2.7}$$

where γ_{sr} is the receding static contact angle. Both empirical constants may differ for the advancing contact line motion ($\gamma_d > \gamma_{sa}$) or for the receding contact line motion

($\gamma_d < \gamma_{sr}$). A contact angle hysteresis is present for $\gamma_{sr} \neq \gamma_{sa}$ meaning that the contact line is assumed static for $\gamma_d \in [\gamma_{sr}, \gamma_{sa}]$. The mobility exponent was assumed to be $m = 3$. Equation (2.7) was further extended by [2] too include evaporation for the investigation of the non-isothermal droplet spreading. The resulting relation is Eq. (2.20) which is central for the present investigation.

Another type of a model for the contact line boundary condition was proposed by [40] for the numerical investigation of the lateral g-jitter effects on free-surface deformation in square containers the following way

$$\frac{\partial h^*}{\partial x^*} = \Gamma^* \left(h^* - h_0^* \right), \, \gamma_d > \pi/2$$

$$\frac{\partial h^*}{\partial x^*} = -\Gamma^* \left(h - h_0 \right), \, \gamma_d \leq \pi/2 \tag{2.8}$$

where h^* is the dimensionless transient height of the interface at the contact line whereas the corresponding dimensionless initial height is assumed to be $h_0^* = 1$. The quantity ∂x^* is the dimensionless horizontal coordinate of the container. Equation (2.8) is valid under the assumptions of $\gamma_s = \pi/2$ and small $h^* - h_0^*$. The quantity $0 \leq \Gamma^* \leq \infty$ is the dimensionless hysteresis parameter. When Γ^* is zero, the contact angle is fixed and the contact line is free to move, whereas when $\Gamma^* = \infty$, the contact line is fixed but the dynamic contact angle is allowed to vary. The dynamic contact angle γ_d can be included in Eq. (2.8) explicitly by defining $\partial h^*/\partial x^* = \cot(\gamma_d)$.

A modification to the above model was introduced by [83] for the numerical investigation of the reorientation of a liquid/gas interface in a right circular cylinder upon step reduction in gravity:

$$\cot(\gamma_d) = \frac{\partial h(t)}{\partial r} = \Gamma \left(h(t) - h(\infty) \right) + \cot(\gamma_s), \, \gamma_d > \pi/2$$

$$\cot(\gamma_d) = \frac{\partial h(t)}{\partial r} = -\Gamma \left(h(t) - h(\infty) \right) + \cot(\gamma_s), \, \gamma_d \leq \pi/2 \tag{2.9}$$

where $h(t)$ and $h(\infty)$ are the dimensional transient height of the interface at the contact line and the corresponding 0g equilibrium position of the contact line, respectively. The dimensional radial coordinate of the cylinder is denoted by r. The inclusion of the term $\cot(\gamma_s)$ allows also static contact angles $\gamma_s \neq \pi/2$. The dynamic contact angle γ_d was included in the model by [83] by defining $\partial h/\partial r = \cot(\gamma_d)$.

The case $\Gamma = 0$ corresponds to the fixed contact angle condition of a free contact line, whereas the condition $\Gamma = \infty$ corresponds to a fixed (pinned) contact line with a free dynamic contact angle. For the intermediate range of Γ, the effective dynamic contact angle varies smoothly with the height of the contact line. Thus besides being a hysteresis parameter Γ is identified as a parameter representing the degree of relative slip at the contact line too.

An important empirical relation for the dynamic contact angle was proposed by [39] as a dependence on the CAPILLARY number **Ca**

$$\frac{\cos \gamma_s - \cos \gamma_d}{\cos \gamma_s + 1} = \tanh(4.96 \, \mathbf{Ca}^{0.702}) \tag{2.10}$$

which for small values of $\mathbf{Ca}$ becomes on the right hand side $\tanh(4.96\,\mathbf{Ca}^{0.702}) \approx 4.96\,\mathbf{Ca}^{0.702}$.

Another empirical relation was proposed by [9]

$$\frac{\cos\gamma_s - \cos\gamma_d}{\cos\gamma_s + 1} = 2\,\mathbf{Ca}^{1/2} \tag{2.11}$$

for which [69] proposed different values of the constants

$$\frac{\cos\gamma_s - \cos\gamma_d}{\cos\gamma_s + 1} = b\,\mathbf{Ca}^{a} \tag{2.12}$$

where $a = 0.54$ and $b = 2.24$ for $\mathbf{Ca} > 10^{-3}$, and $a = 0.42$ and $b = 4.47$ for $\mathbf{Ca} \leq 10^{-3}$.

A very important relation concerning the general behavior the *Hoffman-Voinov-Tanner law*

$$\gamma_d^3 - \gamma_s^3 \cong c_t\,\mathbf{Ca} \tag{2.13}$$

where c_t is a an empirical constant. Equation (2.13) is valid for $\gamma_d \leq 135°$ [42]. The special form of the *Hoffman-Voinov-Tanner law* for completely wetting liquids is usually presented in the form

$$\gamma_d \propto \mathbf{Ca}^{1/3} \tag{2.14}$$

Equation (2.14) was derived through the application of the lubrication approximation by [78] and [77] and it fitted well the experimental data by [36] for $\mathbf{Ca} \leq 0.1$ [42].

The term $(cos\gamma_s - cos\gamma_d)$ in the above relations is associated with the unbalanced surface tension force - the driving force behind the dynamic wetting - which is an essential part of some theoretical models of the contact line motion, [7], [37], [28], [6]. An example of the relation between the dynamic contact angle γ_d and the contact line velocity u_{cl} obtained by one of these models [7], based on the molecular-kinetic theory, is

$$u_{cl} = 2\kappa^0\,\lambda\,\sinh\left[\sigma\,(\cos\gamma_s - \cos\gamma_d)\lambda_0^2/2k_B T\right] \tag{2.15}$$

where k_B is the Boltzmann constant and T is the absolute temperature. The quantities κ^0, the equilibrium frequency of the random molecular displacements occurring within the three-phase zone around the contact line, and λ_0, the average distance of each displacement, are key parameters for the model.

2.3 Phase change at the interface and the contact line.

Interfaces and capillarity were introduced in Sec. (2.1). My problem is essentially an interfacial flow with phase change in the presence of solid boundaries. Quantifying this phase change through the the mass flux at the liquid-vapor interface and the contact line is of paramount importance for the present work since it is the needed for needed for the corresponding boundary conditions.

Generally speaking, the process of a phase change is the transformation of one phase or state into another. In my investigation I am only interested in the transformation between the liquid and the vapor phases. At the conditions of a thermodynamic equilibrium, the temperature at which the phase change occurs, is called the saturation temperature, and the corresponding pressure is the saturation pressure. The relation between the saturation temperature and pressure is the Clausius-Clapeyron equation in the case of an ideal gas which reads

$$\ln\left(\frac{p_1}{p_2}\right) = \frac{\Delta h_{lv}}{R_s}\left(\frac{1}{T_2} - \frac{1}{T_1}\right) \tag{2.16}$$

where the indexes 1 and 2 denote different positions on the saturation curve and R_s is the mass specific gas constant. The latent heat of evaporation Δh_{lv} is assumed to be thereby constant.

There are different models for the calculation of the interfacial mass flux. Among the models two approaches are most commonly used, which differ in relation to the interface and the saturation temperature. The first approach assumes that the mass flux is proportional to the local deviation of the interface temperature T_I from the saturation temperature T_S to the mass flux j, which is known as the Hertz-Knudsen model [41]. The second approach assumes that the interface temperature is equal to the saturation temperature. The mass flux is then proportional to the jump of the conductive heat flux [18]. I present a relation for the mass flux according to both approaches. I use the second one at the interface and the first one at the contact line.

A linearized relation of the interfacial mass flux j according to the the Hertz-Knudsen model is derived [79],[12]

$$j = K^{-1}(T_I - T_S) = \left(\frac{\alpha \rho_v \Delta h_{lv}}{T_S^{3/2}}\right)\left(\frac{M_m}{2\pi R_u}\right)^{\frac{1}{2}}(T_I - T_S) \tag{2.17}$$

where K is the non-equilibrium parameter, ρ_v is the density of the vapor and Δh_{lv} is the latent heat of evaporation. M_m is the molar mass, R_u is the universal gas constant and the quantity α is the so called accommodation coefficient. The above relation is used by [11] and [2] for their investigations.

Based on Eq. (2.17) the interfacial mass flux j_{cl} at the contact line can be written as

$$j_{cl} = K^{-1}(T_{w,cl} - T_S) = \left(\frac{\alpha \rho_v \Delta h_{lv}}{T_S^{3/2}}\right)\left(\frac{M_m}{2\pi R_u}\right)^{\frac{1}{2}}(T_{w,cl} - T_S) \tag{2.18}$$

where $T_{w,cl}$ is the wall temperature at the position of the contact line [2]. The above relation is used for the formulation of the relations by [2] and [35] in the next section and of the contact line boundary conditions in Section 3.3. Thus the evaporative mass flux at the contact line is driven by the local superheat $T_{w,cl} - T_S$.

The interfacial mass flux can computed from the jump of the conductive heat flux [18] as

$$j = -\frac{1}{\Delta h_{lv}} \left[\lambda \, \partial_{\mathbf{n}} T + \lambda_v \, \partial_{\mathbf{n}} T_v \right], \tag{2.19}$$

where $\partial_{\mathbf{n}}$ denotes the normal derivative from the liquid to the vapor.

The above relation Eq. (2.19) is later used in Section 3.3 at the interface since I assume that the temperature at the interface is equal to the saturation temperature by the mathematical modeling of the reorientation. Using Eq. (2.17) under this assumption ($T_I = T_S$) gives a zero interfacial mass flux for which I do not have any proof.

2.4 Contact line boundary conditions with interfacial phase change.

The boundary condition at the contact line in the non-isothermal case with interfacial phase change relates three dynamic quantities - the contact line velocity, the dynamic contact angle and the interfacial mass flux at the contact line or the local superheat at the contact line. The relation includes also some material constants or empirical parameters. So the dynamic wetting gets influenced by interfacial mass transfer.

To much knowledge there is only one relation which has been used as a boundary condition at the contact line for the equations of motion for the investigations of the interface dynamics in problems related to the spontaneous wetting with evaporation/condensation. This is the relation by [2]. Additionally I present the macroscopic relations by [35] and by [63] for complementarity reasons.

The general idea behind the non-isothermal contact line boundary condition is that the influence of the interfacial mass transfer on the wetting is superimposed on the isothermal behavior so that on removal of the mass transfer the contact line should exhibit the already known isothermal behavior described by the relations in Section 2.2.2. So conceptually the non-isothermal relation can be thought of as an extension of the isothermal relations by the inclusion of the interfacial mass flux, as it was done by [2].

2.4.1 Relations for the dynamic contact angle

Based on the mass balance at the contact line when evaporation is present [2] postulated an extension of Eq. (2.7) for the investigation of the spreading of a thin droplet on a heated horizontal solid. By adding a term including the evaporative mass flux from the contact line j_{cl} given by Eq.(2.18) a contact line boundary condition was formulated

$$u_{cl} = -\frac{j_{cl}}{\rho \sin \gamma_d} + \begin{cases} \kappa \left(\gamma_d - \gamma_{sa} \right)^m, & \gamma_d > \gamma_{sa} \\ 0, & \gamma_{sr} \leq \gamma_d \leq \gamma_{sa} \\ \kappa \left(\gamma_d - \gamma_{sr} \right)^m, & \gamma_d < \gamma_{sr} \end{cases} \tag{2.20}$$

where the empirical constants κ and m are given by their nonevaporative values, therefore $m = 3$. The newly formulated expression of Eq. (2.20), relating the the contact line velocity u_{cl}, dynamic contact angle γ_d and the evaporative mass flux from the contact line j_{cl}, has several implications. It implies that in the presence of evaporation the contact line velocity and the liquid velocity at the contact line are not the same. The difference between the two velocities, being related to the mass loss from the contact line due to the evaporation, vanishes only when evaporation is turned off ($j_{cl} = 0$). It also obvious that a variation of γ_d no longer demands the motion of the contact line, even for $u_{cl} = 0$ this can happen when $j_{cl} \neq 0$ i.e., in the presence of evaporation. As also pointed out by [2] the assumed form of Eq. (2.20) can be thought of as a leading order superposition of the effects of (weak) mass loss and the effects of (weak) viscous spreading acting to determine the contact-line motion.

The mathematical form of Eq. (2.20) illustrates the effect of the contact line evaporation on the contact line motion when other effects are neglected. By the advancing contact line the evaporative term has a negative contribution, $-j_{cl}/(\rho \sin \gamma_d)$ <0, in contrast to the positive contribution from the non-evaporative term, $\kappa(\gamma_d - \gamma_{sa})^m$ >0. So the evaporation reduces the absolute value of the advancing contact line velocity, which is positive, compared to the isothermal case. By the receding contact line the evaporative term has still a negative contribution, $-j_{cl}/(\rho \sin \gamma_d)<0$, while the non-evaporative term has a negative contribution too, $\kappa(\gamma_d - \gamma_{sr})^m< 0$. So, the evaporation increases the absolute value of the receding contact line velocity, which is negative, compared to the isothermal case. So generally speaking the contact line evaporation retards the advancing contact line whereas it accelerates the receding contact line in the absence of other effects. Apart from this if evaporation is present, $-j_{cl}/(\rho \sin \gamma_d) \neq 0$, the contact line is no longer required to be static when the dynamic contact angle is within the hysteresis interval as in the isothermal case. Under the given assumptions evaporation would produce a receding contact line motion with $u_{cl} = -j_{cl}/(\rho \sin \gamma_d) < 0$ for $\gamma_{sr} \leq \gamma_d \leq \gamma_{sa}$.

Through the above equation the contact angle γ_{ss} by the stationary contact line $u_{cl} = 0$ can be calculated as

$$\gamma_{ss} \left(\gamma_{ss} - \gamma_{sm} \right)^m = \frac{j_{cl}}{\rho \kappa} \tag{2.21}$$

where $\gamma_{sm} = \gamma_{sa}$ or $\gamma_{sm} = \gamma_{sr}$. The assumption was made $\sin \gamma \approx \gamma$ according to the small-angle theory. As pointed out by [26] and reiterated by [23] small-angle theory holds experimentally up to an angle of $100°$ in agreement with earlier experiments by [36].

Apart from the dimensional form of the dynamic contact angle relation given by Eq. (2.20) several non-dimensional forms were presented by [2] too. For this purpose some characteristic scales and corresponding non-dimensional numbers for the case of thin droplets, and a boundary condition relating the evaporative mass flux j to the local interface temperature were introduced.

As horizontal and vertical scales were taken the initial droplet radius, a_0, and height, $h_0 = a_0 \gamma_0 / 2$, where γ_0, is the initial contact angle. The droplet volume was scaled on $2a_0 h_0$. The contact angle, γ, was scaled on the aspect ratio, $\varepsilon = h_0 / a_0$ which in the case of thin droplets was assumed small. The temperature relative to the saturation temperature was scaled on the wall superheat $\Delta T = T_w - T_S$ between the imposed plate temperature and the saturation temperature. The evaporative mass flux scale was chosen from the energy balance on the interface to balance latent heat of vaporization with heat flux, and was given by $\lambda \Delta T / h_0 \Delta h_{lv}$, where λ is the thermal conductivity in the liquid and Δh_{lv} is the latent heat. Two time scales were chosen: an evaporative time scale $t_{ev} = \rho h_0^2 \Delta h_{lv} / (\lambda \Delta T)$, depending on the evaporation, and also a slow viscous time scale $t_v = h_0^2 / v$ where v is the kinematic viscosity of the liquid. So the ratio of the two time scales $t_v / t_{ev} = \lambda \Delta T / (\varepsilon \rho v \Delta h_{lv}) = \mathbf{E} / \varepsilon$ give the ratio of the EVAPORATION number $\mathbf{E}$ according to its definition in Tab. (3.1) and the aspect ratio.

Thus Eq. (2.20) was non-dimensionalized by introducing a dimensionless time $\tau_v = t / (t_v / \varepsilon)$ too

$$\hat{u}_{cl}(\tau_v) = -\frac{\mathbf{E}_{cl}}{\varepsilon \hat{K} \hat{\gamma}_d(\tau_v)} + \begin{cases} \hat{\kappa} \left(\hat{\gamma}_d(\tau_v) - \hat{\gamma}_{sa} \right)^m , & \hat{\gamma}_d(\tau_v) > \hat{\gamma}_{sa} \\ 0 , & \hat{\gamma}_{sr} \leq \hat{\gamma}_d(\tau_v) \leq \hat{\gamma}_{sa} \\ \hat{\kappa} \left(\hat{\gamma}_d(\tau_v) - \hat{\gamma}_{sr} \right)^m , & \hat{\gamma}_d(\tau_v) < \hat{\gamma}_{sr} \end{cases} \qquad (2.22)$$

where $\hat{u}_{cl}$, $\hat{\kappa}$ and $\hat{\gamma}$ are the dimensionless expressions of the corresponding dimensional quantities in Eq. (2.20). The non-dimensional contact line velocity $\hat{u}_{cl}$ is taken as a derivative of the dimensionless contact line position a/a_0 over the dimensionless time τ_v. The quantity $\hat{K}$ is the non-dimensional expression for K as $\hat{K} = K\lambda / (h_0 L)$. By the derivation of Eq. (2.22) the approximation $\sin \gamma \approx \gamma$ was used since the aspect ratio ε was assumed to be small.

Equation (2.20) can be directly used, as done by [2], for the determination of the dynamic contact angle of steady evaporative menisci where the mass loss due to the evaporation from the interface is exactly balanced by a liquid flow into the droplet away from the contact line. In this case the dynamic contact angles takes a steady value $\hat{\gamma}_{ss}$ calculated from Eq. (2.22) with $\hat{u}_{cl} = 0$

$$\hat{\gamma}_{ss} \left(\hat{\gamma}_{ss} - \hat{\gamma}_{sa} \right)^m = \frac{\mathbf{E}_{cl}}{\varepsilon \hat{K} \hat{\kappa}} \qquad (2.23)$$

Since the right hand side is always positive then $\hat{\gamma}_{ss} - \hat{\gamma}_{sa} > 0$ which means that under the given assumptions evaporation always produces a macroscopic contact angle in the steady case greater than the static contact angle. As noted in [2] the result above is independent of the meniscus shape away from the contact line so it holds for a steady evaporating droplet as well as for the leading edge of a steady evaporative liquid film.

The approach and the results by [2] prompted an investigation of the contact angles in evaporating liquids by [35]. In contrast to [2] the relation for the macroscopic contact angle was not postulated but was derived through the solution the evolution equation of the profile of a thin steady evaporating droplet. The derivation was based on the assumption that the deviation of the macroscopic angle from the microscopic angle is accounted for by processes taking place in the slip region. Within this assumption the microscopic contact angle was considered the same as the static contact angle in a dynamical context. Accordingly the observed variation in the macroscopic angle was entirely attributed to the flow and the physical processes present in the small region near the contact line, where slip is significant. Apart from this a major characteristic of the result is that the mathematical forms of the relations for the macroscopic contact angles differ significantly for non-evaporating and evaporating cases.

The steady case with evaporation from [2] was therefore treated by [35] where the droplet shape and breadth are held constant through a continuous flux of liquid through the base, exactly balancing the mass loss due to the evaporation. The same assumptions by [2] were applied here: two-dimensional (axisymmetric) drop; gravity is neglected; the drop is thin so that the lubrication approximation could be used. Besides this thermocapillary and vapor-recoil effects were excluded. So an evolution equation for the droplet profile was formulated in non-dimensional form with several dimensionless numbers as equation parameters as in [2]: the CAPILLARY number $\mathbf{Ca}$, the EVAPORATION number $\mathbf{E}$, the non-equilibrium parameter $\hat{K}$ and the non-dimensional slip coefficient $\hat{\beta}$. The static contact angle γ_s, taken as identical with the microscopic contact angle, was used for the boundary condition imposed on the evolution equation at the contact line in presence of evaporation. The analysis by [35] required $\mathbf{Ca} \ll 1$, $\beta \ll 1$, and $\hat{K} \gg \hat{\beta}$. The drop was subdivided into three matching regions: a central region, a slip region of width β, and an intermediate region of width $l/|ln\beta|$. In this way the non-dimensional macroscopic contact angle $\hat{\gamma}_d$ was assumed to be equal to slope of the the droplet in the central region at its boundary with the intermediate region. So the profile of the droplet in the central region was presented as a function of the dimensionless parameters and $\hat{\gamma}_d$. Accordingly the profile of the droplet in the slip region was presented as another function of the dimensionless parameters and non-dimensional static (microscopic) contact angle $\hat{\gamma}_s$. Since the slopes given by the two corresponding profile functions do not match each other, an intermediate region was introduced where such a matching procedure is realized. So an non-dimensional expression linking the $\hat{\gamma}_d$ and $\hat{\gamma}_s$ was found. Taking into account the treatment deals with the steady case ($\hat{\gamma}_d = \hat{\gamma}_{ss}$) the expression would has the following form

$$\hat{\gamma}_{ss}^4 = \hat{\gamma}_s^4 + \frac{12\mathbf{CaE}}{\varepsilon^4} \left[\hat{K}^{-1} - W(1)\right] ln\left(\frac{2\hat{\gamma}_s e}{3\hat{\beta}}\right) - \frac{4\mathbf{CaE}}{\varepsilon^4 \hat{\gamma}_{ss}}(c_{11} + c_{12}) \qquad (2.24)$$

where c_{11} and c_{12} are constants. The constant $W(1)$ is value at the contact line of the distribution of the liquid flux into the drop. For small values of $\hat{K}$ Eq. (2.24) becomes

$$\hat{\gamma}_{ss}^4 = \hat{\gamma}_s^4 + \frac{12\mathbf{CaE}}{\varepsilon^4 \hat{K}} ln\left(\frac{\hat{\gamma}_s \hat{K} e}{2\hat{\beta}\hat{\gamma}_{ss}}\right) \qquad (2.25)$$

which was compared to Eq. (2.23) in [2]. Equation (2.25) produced $\hat{\gamma}_{ss} = 0.82$ whereas Eq. (2.23) gave a greater value of $\hat{\gamma}_{ss} = 2.73$. Thus a qualitative conclusion was made by [35] that the contact angle is increased by evaporation which is in agreement with the findings of [2] and the experimental evidence reported there. According to [35] there is a large evaporative loss near the edge, but the fluid in contact with the plate is inhibited from moving, leading to an increase in slope and this description holds whatever model is being used.

A correlation of the contact angle dependence on the wall superheat was provided by [63] using numerical simulations. The contact line behavior of a highly wetting, dielectric liquid (FC-72) droplet under superheated conditions was investigated experimentally and numerically. Unusually high contact angles under superheated conditions for an otherwise highly wetting dielectric liquid was experimentally demonstrated. Despite the high wettability of FC-72 droplets under isothermal conditions, a contact angle of more than $40°$ was observed at a superheat of $15\ °C$. Good agreement between the numerical results and the experiments was shown. The numerical simulations of the microscopic three phase contact line revealed that a competition between evaporation, capillary forces, and van der Waals forces at the contact line drives the phenomena governing the macroscopically observed increased contact angles. The dependence of the contact angle on the wall superheat was modeled using a simple power law curve fit to the numerical data giving the dimensional relation

$$\gamma_d = 17.4(\Delta T_w)^{0.323} \tag{2.26}$$

where the macroscopic contact angle γ_d is given in [$°$] and the dimension of the wall superheat ΔT_w is [$°C$].

2.5 Capillary-driven flow with interfacial phase change.

This section presents the previous results regarding the capillary-driven flow with interfacial phase change - capillary rise and droplet spreading - which are of relevance for my investigation. The interfacial evaporation/condensation has an influence on the capillary flow through the mass balance boundary conditions on the interface and the contact line. In this way the mass flux or the corresponding temperature difference affects the motion of interface and the liquid, which is the principle behind the non-isothermal reorientation. This principle is based on the fact that the mathematical description of the reorientation includes both boundary conditions and the reorientation itself is an example of a capillary-driven flow, too [50]. So there is a base isothermal behavior due to capillary-driven flow in the absence of interfacial phase change expressed by the isothermal values of the characteristics of the interface steady state and dynamics. A non-isothermal behavior due to the interfacial phase change becomes superimposed on the isothermal behavior manifested by the variation of the above characteristics with the variation of the mass flux/temperature difference. Therefore the competition between capillarity and evaporation was pointed out as mechanism behind the observed behavior of the interface [64], [2].

In both cases, capillary rise [64] and droplet spreading [11], [2], [13], an explicit equation of motion of the interface can be formulated and solved instead solving the field equations with the interface as a free boundary, in contrast to the reorientation. Accordingly the influence of the interfacial phase change appears explicitly in the terms of this equation.

However, I must point out that next to their clear relation to the reorientation as capillary-driven flow phenomena both the capillary rise and the droplet spreading have differences from the reorientation. By the capillary rise gravity has to be taken into consideration, too, whereas the droplet spreading occurs under conditions of the lubrication approximation, usually connected with the very low values of the Reynolds number.

2.5.1 Capillary rise with phase change.

By the capillary rise there is an establishment of a steady-state of the liquid-vapor interface in tubes with small radii. Until the steady-state is established there is in the general case a transient unidirectional motion (rise) to the steady-state position followed by damped oscillations about this position. When the oscillations are settled a steady-state interface is observed. The motion of the liquid and the interface is attributed to a capillary-driven flow whereas the the steady-state is considered as the balance of capillarity and gravity in the isothermal case [19]. The dynamics during the damped oscillations can be quantitatively characterized through their frequency and the steady-state of the interface through the height of its position. So both the physical mechanism as well as the characterization of the capillary rise with phase change are directly related to the reorientation. The main results presented here are the summary of the investigations by [64] (numerical), [66] (numerical) and [61] (numerical and experimental). Water and ether were considered as liquids by [64] whereas octane, n-Pentane and iso-octane were studied by [61]. The investigation by [66] was conducted in terms of variation of dimensionless numbers.

As already mentioned an explicit equation of motion of the liquid-vapor interface can be formulated and solved analytically for the transient capillary rise with phase change - the modified Lucas-Washburn equation [64], [66], [61]. The Lucas-Washburn theory was extended by [64] through the addition of interfacial phase change to the capillary flow dynamics. The addition is done through the mass balance boundary condition on the interface accounting for the phase change [64], [66], [61]. This boundary condition reads

$$u_{av} - \frac{dh_{av}}{dt} = u_{av} - u_{I,av} = \frac{j}{\rho} \tag{2.27}$$

where u_{av} is the average axial velocity of the liquid, h_{av} is a cross-sectional averaged interfacial height with respect to the level of the liquid in the reservoir, and j and ρ are the interfacial mass flux and the liquid density, respectively. The quantity $u_{I,av}$ is the velocity of the averaged interfacial height. So the above equation is mathematically identical with the mass balance boundary condition on the interface given by Eq. (3.17) from the mathematical description of the reorientation in Section (3.3.1).

The above equation states that the position/velocity of the interface depends on the rate of interfacial mass transfer relative to the capillary-driven flow. Several features of the flow were emphasized by [64] based on Eq. (2.27). First, when phase change does not take place, zero interfacial mass flux ($j = 0$), the interfacial motion is driven by capillarity and is governed by the standard Lucas-Washburn equation. Next, in the presence of evaporation (condensation), $j > 0$ ($j < 0$), the motion of the interface will be slower (faster) than with no phase change due to the afore-mentioned mass balance boundary condition. These features illustrate the influence of the phase change on the capillary flow - a competition between capillarity and evaporation in contrast to a cooperation between capillarity and condensation.

Through the modified Lucas-Washburn equation analytical expressions were derived by [64] for the steady-state height and of the frequency of the damped oscillations of the interface as functions of mass flux j calculated by Eq. (2.17) using temperature difference across the interface ΔT. The equation of the steady-state height reads

$$h_{ss} = \frac{\dfrac{2\sigma}{\rho R} - \dfrac{j^2}{\rho \rho_v}}{\dfrac{8\mu j}{\rho^2 R^2} + g} \tag{2.28}$$

for the capillary tube with a radius R where the vapor density is ρ_v. The surface tension and dynamic viscosity of the liquid are denoted by σ and μ, respectively. In the case with no interfacial mass transfer $j = 0$ the above equation reduces to

$$h_{ss,i} = \frac{2\sigma}{g\rho R} \tag{2.29}$$

where $h_{ss,i}$ is the equilibrium height under isothermal conditions.

Stability analysis was performed by [64] of the steady-state meniscus for small disturbances to h_{ss} by linearizing the modified Lucas-Washburn equation. Through this analysis an analytic expression for frequency ω of the resulting damped oscillations around h_{ss} was derived

$$\omega = \sqrt{\frac{g}{h_{ss}} - \frac{16\mu^2}{\rho^2 R^4} + \frac{j}{\rho h_{ss}} \left(\frac{8\mu}{\rho R} - \frac{j}{4\rho h_{ss}} \right)} \tag{2.30}$$

in which the influence of interfacial mass transfer is included explicitly through j and implicitly by the dependence of h_{ss} on j. Apparently when no interfacial phase change is present ($h_{ss} = h_{ss,i}$) the above reduces to

$$\omega_i = \sqrt{\frac{g}{h_{ss,i}} - \frac{16\mu^2}{\rho^2 R^4}} \tag{2.31}$$

which gives the frequency of the damped oscillations around the isothermal steady-state height $h_{ss,i}$ from Eq. (2.29).

A variation of the system behavior due to the phase change was reported by [64], [66], [61] with the variation of j. Evaporation was found generally to lower the steady-state height and raise the oscillation frequency. On the other hand, condensation showed the opposite effect leading to a higher steady-state height and to a lower frequency.

Besides the dependence of the results on j, a material dependence was also identified. In the case of [64] for any given capillary radius, ether oscillates faster than water, possibly due to the lower surface tension which results in a smaller amplitude of motion. It is also observed that the water oscillations decay faster due to the stronger damping effect of the higher viscosity of water. In this context the results by [61] showed a clear difference between the results with octane, n-Pentane and iso-octane, too, whereas the material dependence on the properties of the liquid penetrating the capillary tube in the case of [66] was demonstrated through the dependence on a corresponding dimensionless parameter.

A distinct change in velocity was observed experimentally during the early portion of the capillary rise by [61]. This decline in velocity was observed in all heated cases of the $R = 0.5$ mm capillary tests, and most pronounced in the n-pentane and acetone experiments. This result is in accordance with the assessment by [64] following from the mass balance boundary condition of Eq. (2.27) that in the presence of evaporation the motion of the interface will be slower than with no phase change.

Prior to the above investigations by [64], [66], [61] an experimental and theoretical investigation was performed by [80] of the steady-state of the evaporating meniscus formed on a flat plate immersed in a pool of liquid. Four liquid-solid systems were investigated with particular regard to the heat transfer characteristics of the evaporating meniscus: stainless steel-water, aluminum-water, stainless steel-methanol, and aluminum-methanol. A decrease of the meniscus height was observed compared to the isothermal case, demonstrated by a downward migration of the contact line.

The dynamics of an evaporating meniscus on the end of a vertically oriented capillary tube was investigated by [65] as well. He derived equations governing the dynamics of the meniscus by considering an axial, evaporation-driven fluid flow. The motion resulting from a step increase in the evaporation rate was shown to depend on the value of a single dimensionless parameter which is a function of the liquid properties and the dimensions of the capillary tube. A monotone decrease of the curvature radius of the meniscus (corresponding to a increase of the contact angle) occurred for values of the parameter not exceeding a critical value, above which a damped oscillatory motion about a steady state resulted.

2.5.2 Droplet spreading with phase change.

The spreading of small (thin) droplets on a horizontal surface are from special importance for the current investigation. This is a process of spontaneous dynamic spreading of an interface from an initial out-of-equilibrium configuration towards an equilibrium or steady-state configuration by which gravity is neglected due to the small size of the droplet [17]. The motions of the interface and the liquid are essentially capillary-driven or capillary-dominated in the isothermal case [17], [2]. So the spreading of small droplets can be interpreted as an interface reorientation on a small scale with the additional condition of the lubrication approximation which is usually assumed for their investigation.

The effect of the interfacial phase change is incorporated by the inclusion of the mass flux term in the mass balance boundary condition at the interface similar to Eq. (2.27) [2], [35] and in the contact line boundary condition of Eq.(2.20) [2].

The mass flux is calculated through Eq. (2.17) with wall superheat $\Delta T = T_w - T_S$. In this way both the boundary conditions become functions of the wall superheat [60]. From the interface boundary condition an explicit equation of motion of the interface is derived under the assumption of the lubrication approximation [2], [35].

As mentioned above the spreading dynamics of volatile droplets on a heated plate was investigated by [2]. Their aim was to describe the evaporation process macroscopically neglecting the effects of gravity and van der Waals attractions. For the purpose a new contact-line condition was formulated taking into regard both the effects of evaporation and viscous spreading which was presented in Section (2.4.1).

They presented in terms of dimensionless numbers a single differential equation describing the dynamics of the droplet profile for a constant wall superheat within the lubrication approximation. Evaporation enters this equation through the evaporation number $\mathbf{E}$ whereas capillarity through the inverse of the capillary number $\mathbf{Ca}^{-1}$.

Accordingly a quantitative measure of the competition between surface tension and evaporation was defined as a dimensionless parameter which I refer to as the competition parameter C^*

$$C^* = \frac{\varepsilon\,\mathbf{Ca}^{-1}}{\mathbf{E}} \propto \frac{\sigma_0 h_0}{\rho_l\, v^2}\,\frac{h_0/t_v}{h_0/t_{ev}} \propto \frac{\sigma_0 h_0}{\rho_l\, v^2}\,\frac{U_v}{U_{ev}} \tag{2.32}$$

where all the quantities besides U_v and U_{ev} were explained for Eq. (2.22) and $\mathbf{E} = t_v/t_{ev}$ according to [2] was used. U_v and U_{ev} are the viscous and the evaporative velocity scales. So two general cases were identified: $C^* \gg 1$ and $C^* \ll 1$. Accordingly two limits were were considered for the investigation. In the first one, the small capillary number limit $\mathbf{Ca} \to 0$ corresponding to $C^* \gg 1$ and $U_v \gg U_{ev}$, surface surface tension was regarded as the dominant physical mechanism affecting the droplet dynamics whereas in the second case, the large capillary number limit $\mathbf{Ca}^{-1} \to 0$ corresponding to $C^* \ll 1$ and $U_v \ll U_{ev}$, - evaporation.

Through the evolution equation the droplet profiles for two cases were calculated: with (steady profiles) and without (unsteady profiles) a liquid mass source on the solid surface. The steady case means a balance of the mass loss due to evaporation and the mass gain from the liquid mass source. In this way a steady-state value of the contact angle is established which was already given through Eq. (2.23) and is greater than the value without evaporation in the steady-state case of pure spreading. Accordingly the radius of the droplet much is larger with pure spreading whereas the height is much larger with evaporation. Apparently the action of evaporation is opposite to the action of viscous spreading due to capillarity regarding the steady-state characteristics of the contact line and the interface.

Regarding the results with the unsteady profiles three main limits were considered: zero capillary number, small capillary number, and large capillary number. The first two correspond to a dominance of surface tension whereas the last one corresponds to a dominance of evaporation

In the limits of zero and small capillary number two regimes of evaporation were used through a variation of the evaporation number corresponding to a variation of the wall superheat: weak evaporation with $E/\varepsilon = 0.05$ and and strong evaporation with $E/\varepsilon = 0.5$. The the small capillary number limit was realized with $\mathbf{Ca} = 0.1$ leading to $C^* = 200$ for the weak evaporation and $C^* = 20$ for the strong evaporation. These two regimes could be explained in terms of competition between the spreading due to capillarity and evaporation by evaluating the contact line condition Eq. (2.22) at the initial time and neglecting the hysteresis. So the contact line velocity $\hat{u}_{cl}(\tau_v)$ at the $\tau_v = 0$ could be either positive or negative depending on the magnitude of the evaporation number $\mathbf{E}$ according to

$$\hat{u}_{cl}(0) = -\frac{\mathbf{E}}{\varepsilon \hat{K}\hat{\gamma}_d(0)} + \hat{\kappa}\hat{\gamma}_d^m(0) > 0 \ (< 0) \tag{2.33}$$

where $\hat{\gamma}_d(0)$ is the non-dimensional dynamic contact angle at $\tau = 0$.

So although evaporation was always present, $\mathbf{E} > 0$, resulting in a receding motion and eventual disappearance of the droplet, the droplet could still spread initially. And since the evaporation number is proportional to the superheat the condition for the initial spreading or receding would be given by the magnitude of $T_w - T_S$. The above values of $\mathbf{E}$ were so chosen that an initial positive contact velocity (initial spreading) with $\mathbf{E}/\varepsilon = 0.05$ and an initial negative contact velocity (initial receding) with $\mathbf{E} = 0.5$ were ensured. In this way the variation of the evaporation number was manifested by a variation of the contact line condition itself which in turn resulted in a change of the interface dynamics. So in the case of strong evaporation with $\mathbf{E}/\varepsilon = 0.5$ the droplet radius decreased further monotonically in time with a receding motion of the contact line - the maximum value of the drop radius was just the initial value of 1 in non-dimensional terms. In the case of weak evaporation with $\mathbf{E}/\varepsilon = 0.05$ the droplet initial spread through an advancing motion of the contact line reaching maximum radius of slightly under 1.1 followed by a receding motion with a decreasing radius. The case of pure spreading with no evaporation, $\mathbf{E} = 0$, the droplet spread with a monotonically increasing radius to its equilibrium configuration resulting so in a much larger value of 4.5 for the maximum radius. In this sway the opposite actions of and the competition between viscous spreading due to capillarity and evaporation are apparent. The dependence of the interface dynamics on the applied superheat $T_w - T_S$ is so clear.

The dominance of the surface tension and therewith of the viscous spreading could be observed only initially for the case with weak evaporation. However, one must take into account the specific liquid geometry of a thin small droplet (small contact angle) regarding the results - the receding motion of the contact is not only a local dynamic effect but also a consequence of the constantly reduced volume of the whole droplet because of the interfacial evaporation. Nevertheless, the results demonstrate two important things: first, that the dynamics of interface are influenced by the value of the dimensionless parameter C^* and so by the value of the superheat $T_w - T_S$, and second, that the contact line condition can control the behavior of the whole interface and the contact lines can drive bulk motions, as pointed out by [17].

Another characteristic feature of the results was the establishment of a steady-state-like dynamic contact angle due to the balance of spreading and evaporation for the significant part of the life span of droplet. A value with roughly 2.7 in non-dimensional terms is observed in the case of the strong evaporation whereas the value is approximately 1.4 is in the case of weak evaporation. This shows that not only the dynamics of the interface but also the associated with it steady-state can depend on the superheat $T_w - T_S$.

The dynamics of volatile liquid droplets on heated surfaces was investigated by [73] both numerically and experimentally. Instead of a variation of the temperature difference between the liquid and the solid the heating power was varied. It was found that droplet dynamics depends on both heater power and wetting properties of the liquid. The results in this regard are similar to the above discussed results by [2] in the zero and small capillary limit for the weak evaporation regime: an initial capillary spreading-type behavior after which evaporation takes over corresponding to a receding motion of the apparent contact line region [73]. In the initial stage the height of droplet at its center decreases essentially independent of the heating power and the droplet radius increases. After this stage a decrease both in height and radius is observed. Although the droplet ultimately disappears the situation is analogous to a steady-state situation when a finite contact angle is established due to evaporation even for liquids that are perfectly wetting under the isothermal conditions [62]. As pointed out by [73] in the absence of evaporation the droplet would be spreading indefinitely due to capillary forces since the disjoining pressure is inversely proportional to the cube of film thickness. So the results above are another example of the competition between capillarity and evaporation and their opposite action.

A review on the evaporation of macroscopic sessile droplets was made by [13]. A dimensionless parameter was presented, similar to the dimensionless parameter C^* by [2], controlling the dynamics of the contact line and of the evaporating drop by complete wetting. The parameter $\hat{C}^*$ scales like the ratio between the contact line velocity due to evaporation $U_{cl,ev} = J_0/(R\gamma_d)$ and the spontaneous spreading velocity $U_{cl,cap} = \sigma \gamma_d^3/\mu$ which is actually a contact line velocity due to capillarity. J_0 is a evaporation parameter defined as $J_0 = D_m(\rho_{v,int} - \rho_{v,\infty})/\rho$ where D_m is the diffusion coefficient, $\rho_{v,int}$ is the vapor density just above the interface, $\rho_{v,\infty}$ is usual vapor density far way from the interface and ρ is the liquid density. R is the drop radius and γ_d is the dynamic contact angle. The parameter is defined then by

$$\hat{C}^* = \frac{U_{cl,ev}}{U_{cl,cap}} = \frac{3\mu J_0}{\sigma R_{max}\gamma_{max}^4} \tag{2.34}$$

where R_{max} is the maximum radius of the droplet and γ_{max} is the value of corresponding dynamic contact angle. So in principle $\hat{C}^*$ can be also considered a quantitative measure of the competition between evaporation and spreading due to capillarity depending on a evaporation parameter and the material properties.

What is more important that the parameter $\hat{C}^*$ appears directly as a dimensionless number in the equation of motion of the interface evolution derived by [13] in which gravity and thermo-capillary flow are neglected. The equation is presented in a dimensionless form similar to the one derived by [2] for the regime of dominance of evaporation.

2.6 Reorientation and damped interface oscillations

The process of the transition of the macroscopic liquid-vapor interface from its steady-state position under normal gravity to a new steady-state position under reduced gravity is referred to as interface reorientation. It is an example of the spontaneous wetting due to the step reduction of the gravity level. Until the new steady-state position is reached (interface settling) the interface exhibits in the general case decaying damped oscillations. In this context the investigation of damped oscillations of the interface under reduced gravity is related to the reorientation investigation.

In the isothermal case the motion of the interface and the liquid during the reorientation is a manifestation of the capillary-driven flow [50]. The new steady-state position is capillary-dominated and the interface configuration is described solely by the static contact angle [50].

The center point of the interface and the contact line point are used for the quantitative characterization of the reorientation and the damped oscillations [82], [40], [83] and [50]. Accordingly quantitative characteristics of the motion and of the steady-state of these two-points were proposed: deflection of both points (transient, maximal, steady-state); settling time, frequency and damping of the oscillations of the center point; rise velocity and number of oscillations until pinning of the contact line.

The above characteristics depend on material properties and parameters of the liquid-vapor-system, its geometry and gravity levels. This dependence is theoretically expressed through the field equations, the boundary conditions at the interface and the contact line, and through the dimensionless numbers arising from them [82], [40], [83], [50], [29]. Besides this the functional form of the contact line boundary conditions has a significant influence on the interface motion [29],[55].

A summary of the previous investigations on the reorientation of the macroscopic liquid-vapor interface and damped oscillations follows. Dependencies of the characteristics of the center point and the contact line are discussed. The results on the isothermal reorientation and the damped oscillations are presented in Section 2.6.1. No previous results with interfacial phase change are known to me so results on the non-isothermal reorientation with thermo-capillary flow are presented in Section 2.6.2 for completeness.

2.6.1 Isothermal reorientation and interafce oscillations

Experimental investigations

The first known investigation of the reorientation of the liquid-vapor interface was the experimental study by [72] of several storable liquids in different container geometries and diameters D. For the purpose drop tower experiments were performed using partly filled cylindrical, spherical and annular containers. Perfectly wetting liquids were chosen with a static contact angle $\gamma_s = 0°$. The interface motion was found to exhibit characteristics typical of under-damped systems - it oscillates about its new steady-state configuration under reduced gravity.

Due to the limited time of reduced gravity of 2.25 s and the low values of the OHNESORGE number ($0.4 \times 10^{-3} \leq$ **Oh** $\leq 1.6 \times 10^{-3}$ taken with $L = D/2$) a complete decay of the oscillations (total reorientation or settling) was not achieved. That is why an experimental correlation of the time t_{s1} needed by the interface center point to pass its steady-state position under reduced gravity was determined. A functional dependence of t_{s1} on the liquid properties and geometry of the solid-liquid-vapor system (container shape and fill ratio) was established in the form of

$$t_{s1} = K_{emp} \left(\frac{\rho}{\sigma} \right)^{1/2} D^{3/2} \tag{2.35}$$

where K_{emp} is an empirical constant depending on the geometry of the solid-liquid-vapor system, and ρ and σ are liquid density and the surface tension, respectively. A relation to the WEBER number **We** was proposed as $K_{emp} \propto$ **We**$^{-1/2}$. Thereby it was assumed that the characteristic liquid velocity U and the characteristic length L used in the calculation of **We**, defined in Tab.(3.1), are given by $U = t_{s1}/D$ and $L = D$. The effect of viscosity on t_{s1} were neglected by this investigation.

Based on the results in [72] similar drop tower experiments on the reorientation of the liquid-vapor interface osf cryogenic liquids were performed by [71] with cylindrical and spherical containers. Special provisions were made to reduce the heat inputs to the experiment tanks. Thus considering the flight time of 2.25 s, additional sources can be neglected and the reorientation process can be regarded as isothermal. An experimental determination of t_{ss} for the given geometries and liquids was achieved. The data showed good agreement with Eq.2.35 applied to the cryogenic liquids. Thus the extension of the WEBER number scaling to cryogenic liquids was corroborated. It was concluded through the experimental results that the static contact angle γ_s of liquid nitrogen and liquid hydrogen on glass is $0°$. In accordance with this it was also concluded that the configuration of the liquid-vapor interface of nitrogen in a cylindrical tank under reduced gravity is a surface of constant curvature having a radius of curvature equal to the radius of the cylinder.

The earliest quantitative investigation of the characteristics of the interface oscillations after a step reduction of gravity was done by [82]. Data from prop tower experiments with partly filled cylindrical geometries and storable liquids were performed. The primary goal of the investigation was to determine the total reorientation time or settling time t_s as function of the static contact angle and kinematic viscosity. The quantity t_s, defined theoretically as time needed for the complete decay of the oscillation of center point Z_c of the interface, was determined empirically from the temporal evolution of Z_c. A quantitative investigation of the contact line was not performed. Neither a dynamic contact contact relation nor the contact line hysteresis was taken into account by the investigation. The liquids which were used had a static contact angle in the range $0 \leq \gamma_s \leq 74°$. The kinematic viscosity was varied within 6.5×10^{-7} m^2/s $\leq v \leq 5 \times 10^{-4}$ m^2/s resulting in a range of the OHNESORGE number $1.3 \times 10^{-3} <$ **Oh** < 1.1. Through a scale analysis the parameter $\hat{\zeta} =$ **Oh**$\sqrt{\cos \gamma_s}/(1 - \sin \gamma_s)$ (referred to as damping ratio) was identified as the second important parameter besides the static contact angle γ_s to correlate the experimental data. An empirical correlation for t_s was achieved which is essentially a function of γ_s and **Oh** in the form

$$t_s = \frac{R^2}{\nu} \left(1 + \hat{\alpha}^2\right)^{-1} \left(10^B \hat{\zeta}^A + 0.01 \hat{\alpha}^2\right) \tag{2.36}$$

where the parameters a, b and $\hat{\alpha}$ are defined through $a = -1.2\hat{\alpha}^2 + 2.2\hat{\alpha} + 0.28$, $b = 3.9a - 3.32$ and $\hat{\alpha} = (1 - \sin\gamma_s)/\cos\gamma_s$, respectively.

The influences of γ_s and **Oh** on the frequency of the oscillation of Z_c were also presented for $R = 9.52$ mm. A noticeable strong effect of the static contact angle on the oscillation frequency was demonstrated through the increase of the oscillation frequency with an increase in γ_s. For $\gamma_s = 0°$ a pronounced increase of the oscillation frequency with the increase of **Oh** is observed for $\mathbf{Oh} > 3 \times 10^{-3}$. For the bigger static contact angles an increase $32° \leq \gamma_s \leq 68°$ of the oscillation frequency depending on the value of γ_s was shown.

The methodology of [82] was extended in [52], [50] and [53] by including the motion of the contact line Z_c to the quantitative investigation of the interface oscillations, taking place in cylindrical geometries after step reduction of gravity, besides the center point Z_c of the interface. Additionally the influence of the contact line behavior on the oscillation characteristics of Z_c was investigated.

In [50] a mathematical description of the dynamics of the reorientation was developed through the formulation of the corresponding governing equations and the boundary conditions at the interface. No boundary condition was applied at the contact line and the influence of this boundary condition was not investigated. However, the need for an appropriate modeling of the contact line was noted by the author - to give an appropriate boundary condition for the definition of the its motion and for the behavior of the dynamic contact angle. The field equations were non-dimensionalized and dimensionless numbers arose similarly to [40] for the damped interface oscillations. The dependence of the experimental results on these dimensionless numbers was presented and discussed. The interface boundary conditions were not non-dimensionalized, however no other dimensionless number would have been derived [40].

Through specific scaling procedures of the equations the reorientation was subdivided into two time domains. The flow during the initial time domain immediately after the step reduction of gravity is assumed to be independent of the cylinder radius R. Instead the capillary length L_c was assumed as a characteristic length scale. The scaling of the governing equations based on L_c delivered two dimensionless quantities which were identified as characteristic of the flow in this time domain - the MORTON number **Mo** and the acceleration ratio between final and the initial acceleration level. The initial static contact angle, defining the boundary condition for the initial interface configuration, was also assumed as characteristic during this time domain. During the following global time domain the the whole interface becomes involved in the reorientation motion and therefore the cylinder radius R was assumed to be a characteristic length scale in the system. Accordingly the scaling of the governing equations based on R delivered the OHNESORGE number **Oh** (which is formally the MORTON number with L_c substituted by R) and the initial BOND number **Bo**$_i$ as characteristic dimensionless quantities of the flow in this time domain. The static contact angle γ_s, defining the boundary condition of the final configuration of the interface, angle was also assumed to be characteristic in the global time domain. It must be pointed out that the influence of the contact boundary condition, as in [40], or the corresponding dynamic contact angle relation, as in [29] and [55], was not taken into account by the above scale analysis.

An extensive series of experiments was performed by [50] where the characteristic dimensionless numbers were varied according to the above mentioned scaling within these ranges: $0.98 \times 10^{-3} \leq$ **Oh** $\leq 20.17 \times 10^{-3}$, $27.7 \leq$ **Bo**$_i \leq 194.5$, $3.63 \times 10^{-3} \leq$ **Mo** $\leq 52.75 \times 10^{-3}$. The static contact angle was varied within two ranges - $0° \leq \gamma_s < 10°$ and $35° \leq \gamma_s \leq 84°$.

The initial time domain is characterized by the transient capillary rise motion of the contact line. Three characteristics of this motion and their dependencies were analyzed - the dimensionless rise velocity and the dimensionless maximum deflection of the contact line, and the corresponding advancing dynamic contact angle. It was found that the dimensionless rise velocity was mainly dependent on theMORTON number and by the static contact angle for big values of the initial BOND number (**Bo**$_i \geq 100$). An increase of the velocity with decreasing Mo and γ_s values was shown. For the smaller values of **Bo**$_i < 100$ there was an influence of **Bo**$_i$ on the rise velocity whereby the velocity decreased with the decrease of **Bo**$_i$.

The dimensionless maximum deflection of the contact line was found to be dependent on the OHNESORGE number, the static contact angle and the initial BOND number in different ways. It decreased with the increase of **Oh**, decreased with the increase of static contact angle and increased with the increase of **Bo**$_i$. Besides this a different motion behavior of the contact line for small ($\gamma_s < 10°$) and big ($\gamma_s \geq 35°$) static contact angle was observed.

For $\gamma_s \geq 35°$ the contact line exhibited an overshoot above its equilibrium position under reduced gravity followed by an oscillating motion with a advancing and receding contact dynamic contact angles for the majority of the experiments. The overshoot decreased with the increase of **Oh** until no overshoot was observed for **Oh** $\geq 20 \times 10^{-3}$ when the contact line was immediately anchored. The oscillations decayed within a few cycles and an anchoring of the contact line due to the hysteresis of the dynamic contact angle was observed after which it remained fixed while the center point of the interface continued to oscillate until settling occurred. This behavior had been reported also by [82]. The number of the contact line oscillations until the anchoring and the dimensionless time of the anchoring decreased with an increase of **Oh** and of γ_s.

A very different behavior of the contact line was reported for $\gamma_s < 10°$. This difference was found to be very important for the description of the oscillation of the center point. An overshoot was reported only for a small minority of experiments with **Oh** $\leq 1. \times 10^{-3} - 2.5 \times 10^{-3}$ and this overshoot decreased with the increase of **Oh**. However, no receding motion of the contact line for these cases was observed, instead a macroscopic liquid layer was formed on the wall. This layer drained during the continuing reorientation and in the transition region between the layer and the oscillating interface an apparent contact point was formed. The motion of this apparent contact point along the cylinder wall can be approximated by the boundary condition of free contact line with a fixed contact angle in [67]. For $\mathbf{Oh} \geq \times 10^{-3} - 2.5 \times 10^{-3}$ there was no overshoot of the contact line above final equilibrium position and the contact line moves towards its final equilibrium position in an very slowly. This advancing motion of the contact line can be approximated by the boundary condition of a fixed contact line with a free contact angle in [67]. In this case the interface oscillated around the contact line position beneath the final contact line position which position corresponds to a higher contact angle than the static contact angle of the liquid.

The dynamic contact angle during the transient capillary rise was also performed within a range of the CAPILLARY number $1.1 \times 10^{-4} \leq \mathbf{Ca} \leq 4.5 \times 10^{-3}$ where $\mathbf{Ca}$ was calculated through the rise velocity of the contact line. The data was compared to the published experimental data from [36] and [76], to the empirical correlations of the dynamic contact angles by Eq. (2.10) in [39], by Eq. (2.11) in [9] and by Eq. (2.12) in [69], and to the theoretical model in [27]. It was found that the values were slightly higher than the previous experimental values which are measured mostly under stationary conditions. A very good agreement between the data in [50] with the empirical correlation by [9] was established.

The following global time domain was characterized through the oscillatory motion of the center point Z_c of the interface. Mainly through two characteristics of this motion were investigated - the frequency and the damping of the oscillations. Besides this a dependence of the maximal deflection of the center point on the OHNE-SORGE number and the static contact angle was observed - it increased with the decrease of $\mathbf{Oh}$ and γ_s. The oscillation of Z_c was found to be essentially non-linear as long as the contact line Z_w exhibited an observable motion. After the anchoring of Z_w the oscillation of Z_c could to be assumed as linear, expressed by approximately constant frequency and damping, for $\gamma_s \geq 35°$. For $\gamma_s < 10°$ a linear oscillation could be assumed already after the maximum deflection of the contact line. However for this case the dependence on the behavior of the contact line and therefore on OHNE-SORGE number was taken into account. In this way the frequency and the damping of the approximately linear part of the oscillation of Z_c were investigated.

The mean frequency of the oscillation ω_d was non-dimensionalized with the characteristic time t_{pu} to give a dimensionless mean frequency of the damped oscillations

$$\Omega_d = \omega_d t_{pu} \tag{2.37}$$

where the time $t_{pu} = \sqrt{\rho R^3 / \sigma}$ can be regarded as a characteristic time scale for the natural frequencies of the free oscillations of liquid droplets, [47]. Different dependencies of Ω_d were observed. It increased with the increase of the static contact angle. For $\gamma_s \geq 35°$ this dependence was the only clear influence on the dimensionless frequency by the experiment parameters since no unambiguous influence by $\mathbf{Oh}$ and $\mathbf{Bo}_i$ could be identified. A conversion of Ω_d into a natural frequency Ω_n and an extrapolation of Ω_n for very small OHNESORGE numbers with $\gamma_s = 0°$ delivered a theoretical minimum of $\Omega_n = 3.2$ according to the author.

For $\gamma_s < 10°$ dependencies on both **Oh** and **Bo**$_i$ of Ω_d were observed too. The dimensionless frequency increased with the increase of **Oh** and decreased with the increase of **Bo**$_i$. The increase of Ω_d with the OHNESORGE number was attributed to the slowly creeping motion of the contact line. For small OHNESORGE numbers **Oh** $\leq 1.5 \times 10^{-3} - 2.5 \times 10^{-3}$ the values of Ω_n were close to the assumed minimum value of $\Omega_n = 3.2$. Values of Ω_n below the value of 3.2 were not expected.

An investigation of the oscillations during the reorientation the liquid-vapor interface of a cryogenic liquid was performed by [75] with a cylindrical geometry with an inside diameter of ø 5.1 cm after a step reduction of gravity under isothermal conditions through drop tower experiments. Liquid nitrogen was used as an experimental liquid. Apart from the experiments numerical simulations of the reorientation of liquid nitrogen with the commercial CFD package FLOW3D were also performed. A comparison was presented between the values of some of the reorientation characteristics from the cryogenic experiments, from extrapolations from the storable liquids used by [50] and from the simulations. Thus the time of the maximum deflection of the center point of the interface from the experiments was reported to be $t_{cp} \approx 0.9$ s which is smaller than the value from the simulations $t_{cp} \approx 1.2$ s and significantly smaller than the extrapolated value $t_{cp} \approx 1.5$ s from the storable liquids. The value of maximum deflection of contact line $Z_{wp} \approx 26$ mm from the simulation was reported to be bigger than the extrapolated value from the experiments by [50] whereas no value was reported from the cryogenic experiments. A macroscopic liquid layer was observed on the cylinder wall both in the experiments and in the simulations with liquid nitrogen in agreement with the findings of [50] for $\gamma_s < 10°$ and **Oh** $\leq 1.5 \times 10^{-3} - 2.5 \times 10^{-3}$. Only a uni-directional motion of the contact line could be observed in the experiments.

Numerical and analytical investigations

A study by [4] investigated analytically and numerically the frequency behavior of an interface which due to small disturbances oscillates about its steady-state (equilibrium) configuration under zero gravity in a partly filled cylinder. Accordingly the motion of the interface from its equilibrium position under normal gravity was not considered. Several assumptions were made for the treatment of the problem. The liquid was considered frictionless leading to the existence of a velocity potential. A free contact line with a fixed contact angle on the wall was assumed. The amplitude of interface oscillations about a spherical (steady-state (equilibrium) shape under zero gravity was considered small. Under the above assumptions the LAPLACE equation was solved for which the kinematic (mass-balance) and the dynamic boundary conditions on the interface were linearized. The application of the BESSEL function of the first kind yields expressions for the solution of the LAPLACE equation and for the interface elevation. The substitution of these expressions into the linearized kinematic and dynamic surface boundary conditions renders an eigenvalue problem. Through the determination of the eigenvalues λ_{mn} the natural frequencies ω_{mn} of the interface oscillations are obtained as functions of the liquid properties and the geometry of the system

$$\omega_{mn}^2 = \frac{\sigma}{\rho R^3} \lambda_{mn} \tag{2.38}$$

where m and n denote the corresponding lateral and axial oscillation mode, σ and ρ are the surface tension and the density of the liquid, and R is the cylinder radius. The interface oscillations during the reorientation correspond to the first axial mode with ω_{01}. The influence of the dimensionless static contact angle $0.2 \leq \gamma_s/\pi \leq 0.8$ and the fill ratio $0 \leq h/R \leq 1$ (where h is fill height of the liquid) on the dimensionless natural frequency $\omega_{mn}/\sqrt{\sigma/\rho R^3}$ was investigated by determining ω_{mn} numerically. Here it must be pointed out that $\omega_{mn}/\sqrt{\sigma/\rho R^3}$ is equal to the frequency Ω_d from Eq. (2.37) where $t_{pu} = \sqrt{\rho R^3/\sigma}$ taking into account also that $\Omega_d = \Omega_n$ due to the lack of damping in the system. For $\gamma_s = \pi/2$ an analytical expression of the dependence of Eq. (2.38) on the fill ratio was presented

$$\omega_{mn}^2 = \frac{\sigma}{\rho R^3} \varepsilon_{mn}^3 \tanh(\varepsilon_{mn} h/R) \tag{2.39}$$

where ε_{mn} are the zeros of the first derivatives of the BESSEL functions of the first kind.

The dimensionless natural frequency $\omega_{01}/\sqrt{\sigma/\rho R^3}$ of the first axial mode exhibited the smallest values of all axial modes. This means that all other axial modes would be much faster damped out when viscous damping is present in the system. Therefore the reorientation oscillations of the interface in a viscous liquid (which are the subject of the investigation of the present work) would be dominated by the ω_{01} mode. Apart from this $\omega_{01}/\sqrt{\sigma/\rho R^3}$ showed a dependence on both the static contact angle and the fill ratio in the investigation by [4]. For $h/R = 0.5$ it showed an increasing trend with the increase of γ_s, reaching a maximum at around $\gamma_s = \pi/2$ and the decreasing for the bigger values of the contact angle. With an increase of the small fill ratios it increased rapidly towards an asymptotic value. The value of Eq. (2.39) at $h/R = 0.6$ is 99 percent of the value at $h/R \to \infty$. Thus the influence of the bottom of the cylinder can be considered as negligible for $h/R > 0.6$.

Damped interface oscillations were numerically investigated with FLOW3D by [40] by a variation of the conditions at the contact line. The range of small amplitude oscillations of the interface in a partially filled rectangular container was implemented. The goal was to capture the influence of these conditions in the isothermal case based on the fact that the contact line behavior significantly affects the dynamics of the interface. For the purpose the corresponding field equations with their boundary conditions were non-dimensionalized. Equation (2.8) was implemented as a non-dimensionalized boundary condition at the contact line. This gave a full equation system describing the fluid motion in a uniformly non-dimensionalized form. In the absence of gravity this equation system shows that the fluid motion depends only on the values of two dimensionless numbers: the Ohnesorge number **Oh**, arising from the Navier-Stokes equation and the normal stress boundary condition at the interface, and Γ^* arising from the contact line boundary condition. Both **Oh** and Γ^* were varied for the numerical investigation. This investigation concept by [40] is used in the present work by extending it to the non-isothermal case with interfacial phase change - the variation of the of axial wall temperature gradients is related to the variation of dimensionless numbers from the interface and the contact line.

The dependence of the resonant frequency on **Oh** was presented for the cases with a free contact line $\Gamma^* = 0$ and fixed contact line $\Gamma^* = 1000$. The value of frequency for the fixed contact line is about as twice as big as the frequency for the free contact line for a given value of **Oh**. When **Oh** becomes larger than about 0.15 for $\Gamma^* = 0$ and 0.3 for $\Gamma^* = 1000$, the viscous effect from the walls becomes dominant, so that the frequency to decrease significantly with the increase of the **Oh**. On the other hand, when Oh is less than about 0.01, the wall effect becomes small and the resonant frequency becomes equal to that for inviscid flow (**Oh** = 0).

The dimensionless hysteresis parameter Γ^* was varied by [40] within the range $0 - 10000$. Results were shown for **Oh** = 0.01 (nearly inviscid flow). A clear dependence of the motion of the interface on value of Γ^* was observed. The dimensionless resonant frequency of the interface oscillation up to about $\Gamma^* = 0.01$ remains equal to that for the free contact line; the resonant frequency then increases with Γ^* up to about $\Gamma^* = 1000$, beyond which it is equal to that for the fixed-contact-line condition ($\Gamma^* = \infty$).

The whole reorientation motion of the interface from its steady-state position from normal gravity was numerically investigated by [83] applying the mathematical description by [40]. However, the equation system was not non-dimensionalized. The commercial CFD package FIDAP was used for purpose.

The dynamic contact angle relation of Eq. (2.9) was implemented as a boundary condition on the contact line in which the parameter Γ was employed as an adjustable parameter. The hysteresis of the contact angle was taken into account. The principal objective of the investigation was to determine the characteristics of the oscillations, in particular the oscillation frequency and the settling time, as functions of Γ for systems of various values of **Oh** and γ_s in analogy to [14] and [40].

The investigation was performed for the cylinder geometry with a radius $R = 9.25$mm and for 6 silicone liquids used in [82]. The OHNESORGE number and the static contact angle were varied accordingly within the corresponding ranges: $0.00146 \leq$ **Oh** ≤ 0.0440 and $32° \leq \gamma_s \leq 66°$. So the liquids studied by [83] exhibited in the experiments by [82] the behavior of underdamped systems with damped oscillations of the interface about its low-g steady-state configuration.

In the numerical simulations Γ was varied analogous to [40] as a free parameter within the range $0 - 1000$ cm^{-1} for combinations of the above mentioned values of **Oh** and γ_s. In this manner the influence of Γ on the dimensional resonant oscillation frequency and the settling time was investigated. For the settling time t_s a quantitative definition was introduced: t_s is defined as the time required for the center point oscillation to decay to ± 2 % of its steady-state deflection. Besides the dependence on Γ the dependencies on **Oh** and γ_s were investigated for fixed values of Γ. For the free contact line condition $\Gamma = 0$ both the frequency and the settling time had the smallest values. With the increase of Γ both the frequency the settling time increased so that for the fixed contact line condition $\Gamma \to \infty$ a constant value is approached - the influence of Γ becomes weak for $\Gamma \geq 100 \, \mathrm{cm}^{-1}$. Both the resonant frequency and the settling time increased with the increase of γ_s and with the decrease of **Oh**. The results are similar to these reported by [82]. On the other side t_s shows a weaker dependence on γ_s in contrast to the frequency.

Apart from this the influence of the fill level of the liquid was investigated. The liquid velocity at depths $1.5R$ was found be only 1 % of the liquid velocity at the interface. Thus the influence of the bottom was not anticipated for fill levels above $1.5R$.

The results of very comprehensive numerical studies of the reorientation behavior in partly filled cylinders were presented in [29] and [30]. The commercial CFD package FIDAP was used for the investigations. Two models for the boundary conditions at the contact line were implemented in [29]. A slip boundary condition was imposed on the wall in vicinity of the contact line which allowed a free slip between the liquid and the wall in a small region around the contact line. For the boundary condition regarding the dynamic contact angle a relation was developed and implemented taking into account the different dependences of the dynamic contact during the reorientation process. The advancing and the receding contact angle were modeled as functions of the CAPILLARY number **Ca** through modifications of the empirical correlation Eq. (2.10) by [39]. Additionally a model of the hysteresis of the contact angle was implemented as a function of the CAPILLARY number and the height of the contact line. Through the two boundary condition models the whole reorientation of the interface from its 1g equilibrium shape was numerically investigated. However, an investigation of the dependence of the reorientation characteristics on **Ca** was not performed.

Besides this the interface oscillations about the low-g equilibrium configuration were studied. In this case the calculations started with the low-g equilibrium configuration. This configuration was disturbed through an short impulse of an axial acceleration, followed by the damped oscillation of the interface about its equilibrium position. On the one hand the contact line could move freely with the the free slip condition $\beta = 0$ valid within a small length from the contact line and with a fixed contact angle $\gamma_d = \gamma_s$. On the other hand the contact line was kept fixed while contact angle could change freely.

The models for the boundary conditions at the contact line were initially validated through a comparison with the experimental results by [82] and [50]. The comparison showed a very good agreement regarding the motions of both the center point and the contact line. In contrast to [83] not only the frequency but also the damping and the amplitude of the center point oscillation were very good reproduced. Considering the contact line motion even the number of extremal points was in good agreement besides the its deflection. The numerical results revealed that the hysteresis of the contact angle had a significant influence on the behavior of the contact line. The hysteresis was found responsible for the anchoring of the contact line in the new equilibrium position under low-g conditions. The fundamental importance of the boundary conditions at the contact line for the reorientation was demonstrated by the very good agreement between the experimental and the numerical data by [29].

In the consequent numerical simulations of the reorientation the relevant dimensionless numbers according to [50] were varied in [29] within the following ranges: $0 \leq \mathbf{Bo}_i \leq 2.5 \times 10^3$, $2.6 \times 10^{-4} \leq \mathbf{Mo} \leq 1.11 \times 10^{-1}$, $1.0 \times 10^{-4} \leq \mathbf{Oh} \leq 5.0 \times 10^{-1}$ and $5° \leq \gamma_s \leq 90°$. In this way the ranges of these dimensionless numbers for the experiments by [50] were substantially extended. So the reorientation behavior could be studied from an almost inviscid flow to a creeping flow for the range of γ_s. Due to the limitations of the finite elements method employed by FIDAP static angles $\gamma_s < 5°$ could not be investigated. Besides different ratios of the liquid fill level h_0 to the cylinder radius R were considered: $h_0/R > 1$ and $h_0/R < 1$.

Through the validated models the reorientation characteristics defined and investigated experimentally in [50] were investigated numerically. In the initial time domain the dependences of the velocity and the maximal deflections of the interface were ascertained. The dimensionless maximum velocity increased with the increase of the initial BOND number and with the decrease of the static contact angle. For $\mathbf{Bo}_i \geq 200$ the increase became much weaker and for $\mathbf{Bo}_i \to \infty$ a maximum value was approached. For $\gamma_s \to 0°$ a maximal value was also approached. The increase of the MORTON number resulted in a decrease of the maximal velocity of the contact line due to the dissipative effects. The mean velocity of the contact line exhibited a qualitatively similar behavior. The flow for the smallest value of the MORTON number could be considered almost frictionless. The potential energy due to the curvature gradients along the interface in the vicinity of the wall was transformed almost without dissipative losses was transformed into biggest possible contact line velocity. The dimensionless maximal velocity of the contact line depends particularly on the initial interface curvature near the wall which curvature is determined by the initial BOND number and the static contact angle.

The maximal deflections of the contact line and of the center point were found to depend on the OHNESORGE number and the static contact angle, and on the initial BOND number for $\mathbf{Bo}_i \leq 200$. Both maximal deflections decreased with the increase of $\mathbf{Oh}$ and γ_s. The flow pattern was dominated by the convective impulse transport for $\mathbf{Oh} \leq 10^{-3}$ whereas for the higher values of $\mathbf{Oh}$ a dominance of the viscous impulse transport was determined. The overshoot of the contact line over its low-g equilibrium position decreased with the increase of OHNESORGE number and for $\mathbf{Oh} \geq 10^{-2}$ the interface approached asymptotically the final configuration without an overshoot.

In the following global time domain the oscillation of the center point could be approximated by the linear oscillator. Both the frequency and the damping behavior were studied in dependence on the boundary conditions on the contact line and on certain viscous effects.

The frequency behavior depended strongly on the imposed boundary condition at the contact line. The lower and the upper limits of the frequency were given by the free contact line (with a fixed contact angle) and the fixed contact line (with a free contact angle) conditions. The dimensionless natural frequency Ω_n of the center point, defined in [50], increased with the increase of the static contact angle whereas the frequency with the fixed contact line was higher than the frequency with a free contact line. For the smallest value of the OHNESORGE number the frequency behavior corresponded to a frictionless flow within the range of the static contact angle. For this case analytical approximations were developed based on [4] which reproduced the frequency behavior of very good. The non-linear region of the dependence of the frequency on the contact angle is expressed by

$$\Omega_{n,free}(\gamma_s) = \varepsilon_{01}^{3/2} \sin^{3/5}(\gamma_s) \, , \, 40° \leq \gamma_s \leq 90° \tag{2.40}$$

and

$$\Omega_{n,fixed}(\gamma_s) = \sqrt{2}\,\varepsilon_{01}^{3/2} \sin^{3/5}(\gamma_s) \, , \, 20° \leq \gamma_s \leq 90° \tag{2.41}$$

where $\Omega_{n,free}$ and $\Omega_{n,fixed}$ are the dimensionless frequencies with a free and with a fixed contact line, respectively, and ε_{01} is the is the first zero of the Bessel function of first kind. The apparent lack of dependence on the OHNESORGE number of the frequency in Eq. (2.40 - 2.41) matched the numerical results for $\mathbf{Oh} \leq 10^{-3}$. This behavior is known as an asymptotic behavior for vanishing OHNESORGE numbers. The limit values of the dimensionless frequency for $\gamma_s \to 90°$ were given by $\Omega_n = 7.5$ with the free contact line and $\Omega_n = \sqrt{2} \cdot 7.5$ with the fixed contact line.

The linear region of the dependence of the frequency on the contact angle is expressed by

$$\Omega_{n,free}(\gamma_s) = 3 + 0.07\frac{\gamma_s}{[1°]} \, , \, 15° \leq \gamma_s \leq 40° \tag{2.42}$$

and

$$\Omega_{n,fixed}(\gamma_s) = 3.5 + 0.13\frac{\gamma_s}{[1°]} \, , \, 10° \leq \gamma_s \leq 20° \tag{2.43}$$

which is very important for the present investigation. The extrapolation of the analytical approximation Eq.(2.42-2.43) for $\gamma_s \to 0°$ in $\Omega_n = 3.0$ with the free contact line and $\Omega_n = 3.5$ and with the fixed contact line.

The above relations both for the non-linear and for the linear region express a monotonic dependence of frequency on the static contact angle. The frequency increases with the increase of the contact angle. So any mechanism which increases the contact, e.g. evaporation, would lead to an increase of the frequency.

With the increase of the OHNESORGE number as $\mathbf{Oh} > 10^{-3}$ the frequency was found to depart from the asymptotic behavior due to two damping effects: the damping due to the viscous boundary layer at the wall and the wave damping within the oscillating bulk flow. The boundary layer effects lead to a blocked region for the flow near the wall because of which the interface near the wall was not involved in the oscillations. This caused an increase of the frequency due to the decrease of the effective diameter of the oscillating interface. The dissipative effects due to the bulk flow resulted in a decrease of the frequency with the increase of the OHNESORGE number. The dissipative effect due to the viscous boundary layer at the wall became apparent with the free contact line for smaller values of **Oh** compared to the fixed contact line. For the very high values of OHNESORGE number the damping was so high that the motion of the center point exhibited a transition from a periodic to a aperiodic behavior.

The dimensionless settling (or total reorientation) time was evaluated from the values of Ω_0 and $\hat{D}$ for both the free and the fixed contact line through the relation

$$t_s/t_{pu} = \frac{\ln 0.02 + \ln \sqrt{1 - \hat{D}^2}}{-\hat{D}\Omega_0} \qquad (2.44)$$

where t_s is the dimensional settling time (see Eq. (2.36)) and t_{pu} is the characteristic time scale in Eq. (2.37). Thereby the deflection of the center point is required to be within 2 % of the end position for $t \geq t_s$. The dimensionless settling time decreased with the increase of the OHNESORGE number and approached a minimal value in the transition region between the periodic and the aperiodic behavior of the center point. An increase of t_s/t_{pu} was observed with the decrease of the static contact, angle attributed to the the increase of $\hat{D}$. The settling time was smaller for the free contact line than for the fixed contact line due to the higher damping in the first case.

Numerical investigation of the reorientation was presented in [56] and [55]. For the investigation a finite volume method was used for the solution of the equations of motion. The position of the free liquid surface has was described with an algorithm based on the volume-of-fluid method by [34]. The numerical scheme followed a mass-conservative displacement algorithm. The goal of the investigations was to study the influence of the boundary condition of the dynamic contact contact angle on the liquid dynamics during the reorientation, in particular on the the motion of the contact line and of the center point of the interface. The relations for the dynamic contact angle which were tested in [55] included a constant dynamic contact angle $\gamma_d = \gamma_s = $ const, the theoretical model Eq. (2.15) based on the molecular kinetic theory and three empirical correlations - Eq. (2.10), Eq. (2.11) and Eq. (2.12). Two cases were calculated for which experimental results were available in [50], [51] and a comparison between simulation and experiment was presented. Two liquids were used in the calculations - detra and M3. Detra has a small static contact angle $\gamma_s = 5.5°$, wheras M3 a large static contact angle $\gamma_s = 53.6°$. Both liquids had values of the initial BOND number $\mathbf{Bo}_i < 75$ and of the OHNESORGE number $\mathbf{Oh} < 10^{-2}$. The experiments with detra exhibited a creeping motion of the contact line without anchoring, whereas the experiments with M3 revealed an oscillating contact line with a quick anchoring.

The comparison between the experimental and the numerical results revealed that the model of the dynamic contact angle could have a large influence on the liquid dynamics in capillary-dominated systems. The first model with the constant dynamic contact angle was found to be highly inaccurate. This model resulted in too violent reaction of the liquid, exaggerated deflections of the center point and the contact line, and the oscillations lasted too long. In contrast to that the use of a model for the dynamic contact angle showed clearly more realistic results. The theoretical model given by Eq. (2.15) was found to work quite well for both the small and the large static contact angle. In the former case a hysteresis of the contact line was not required. The oscillating case with large static contact angle required the addition of hysteresis. Doing so, with experimentally obtained hysteresis angles, the flow dynamic were captured accurately. The empirical correlations could be applied only for the advancing contact line motion. They were found less accurate than the theoretical model.

2.6.2 Non-isothermal reorientation

The first direct investigation of the interface oscillations during the reorientation in cylindrical geometries under non-isothermal conditions were the experiments performed by [31]. However, the thermal effect which influenced the reorientation was the thermo-capillary convection not interfacial phase change. The experiments focused on the investigation of the influence of the non-isothermal boundary conditions on the interface reorientation. For that purpose storable and completely wettable liquids ($\gamma_s = 0$) were used for small values of the OHNESORGE number in cylinder geometries. Through heating the the wall above the interface a temperature difference $T_w - T$ between the wall with a temperature T_w and the liquid with a temperature T was imposed. The volume above the interface was occupied by the vapor and a non-condensable gas, allowing for thermo-capillary convection in the liquid. In this manner due to the temperature difference $T_w - T$ a temperature gradient of surface tension σ_T appears and thermo-capillary flow occurs, described by the thermocapillary REYNOLDS number $\mathbf{Re}_M = \sigma_T (T_w - T) R / (\rho_0 v^2)$ where ρ_0 is mean liquid density, R is the cylinder radius and v is the kinematic liquid density.

The dependence of the final steady-state contact angle γ_{ss} on the thermocapillary REYNOLDS number $\mathbf{Re}_M$ was presented. The dependence of the dimensionless natural frequency Ω_n on γ_{ss} was shown in comparison with numerical data calculated with FIDAP under isothermal conditions for a free and fixed contact line condition.

At the non-isothermal conditions the thermo-capillary flow induces to flow away from the contact line and causes an increase of the apparent contact angle. In this way the initial static contact angle γ_s was shifted to higher values of γ_{ss}. The final steady-state contact angle γ_{ss} increased with the increase of the thermocapillary REYNOLDS number. On the other hand the frequency Ω_n increased with the increase of γ_{ss} and it followed the behavior of the free contact line in the isothermal cases. The damping ratio was reported to decrease with an increasing thermocapillary REYNOLDS number. Besides this an oscillating contact line was observed.

2.7 Synopsis and goals of the present work

The isothermal reorientation is considered to be capillary-driven [50], [29]. Therefore the non-isothermal reorientation can be regarded as an example of a capillary-driven flow with interfacial phase change similar to the cases presented in Section

(2.5). Accordingly the variation of the dynamics and the steady-state of the interface can be expected to be functions of the variation of the mass flux at the interface and at the contact line. Relevant dimensionless numbers arising from the boundary conditions at the interface and the contact line would describe the variation of the reorientation characteristics of the interface.

The general topic of the present work is the influence of the mass and heat transport due to the wall superheat on the linear impulse transport during the reorientation. Therefore two specific major goals were formulated concerning of the interface behavior during the reorientation of single species, two-phase, cryogenic liquids in partially filled containers under diabatic wall boundary conditions.

The first one was the development of a formalism of the reorientation expressing the influence of the phase change at the interface and the contact line. For that purpose a mathematical description of the problem of the capillary driven flow in the presence of interfacial phase change and a superheated wall was formulated. Therefore the treatment used by [50] and [29] was extended to the case of two-phase flow of cryogenic liquids with diabatic walls including heat and mass transfer in the boundary conditions at the interface, and including a corresponding boundary condition at the contact line. This is of primary importance since the unique character imparted to the flow in the presence of fluid and solid boundaries enters the fluid mechanical description as boundary conditions [81]. Accordingly the flow is expected to be influenced by these boundary conditions, as this was already presented in this chapter. Through the formulation of the field equations and boundary conditions the relevant dimensionless numbers are derived, describing how the evaporation at the interface and the contact line affects the capillary-driven flow, which are defined in Table (3.1) in Chapter (3). These dimensionless numbers are the EVAPORATION number E at the interface and E_{cl} at the contact line, and the EVAPORATION heat flux number in the vapor Ev_v. The evaporation number E appears from the non-dimensionalization of the mass balance boundary condition at the interface (Eq. (3.63)), whereas E_{cl} - from the contact line boundary condition (Eq. (3.45-3.46)). The non-dimensional mass flux on the interface (Eq.(3.74)), which enters the non-dimensionalized mass balance boundary condition at the interface, depends on Ev_v. However, both E and E_{cl} appear as products with the OHNESORGE number Oh in the mentioned boundary conditions - $E\,Oh$ and $E_{cl}\,Oh$, respectively. $E\,Oh$ and $E_{cl}\,Oh$ have the meaning of a competition parameter, a measure of the competition (interaction) between capillarity and evaporation, Eq. (3.67). Therefore I regard $E\,Oh$ and $E_{cl}\,Oh$ as the representative dimensionless numbers instead of E and E_{cl} alone.

The second goal was the experimental quantification of the influence of the interfacial phase change (due to the variation of the boundary conditions at the interface and the contact line), and particularly of evaporation, on the interface characteristics, such as deflection, frequency and velocity. For that purpose drop tower experiments were executed and a variation of the axial wall gradients with different liquids was performed. The dependence of the interface characteristics from the experiments on the above mentioned dimensionless numbers is presented in Chapter (5).

Regarding the isothermal behavior this work is an extension of the investigation by [50] of the reorientation of storable liquids with a range of the OHNESORGE number $\mathbf{Oh} \geq 10^{-3}$ to the region of the cryogenic liquids with $\mathbf{Oh} < 10^{-3}$, Table A.11-A.15. The OHNESORGE number was in the range $2.2 \times 10^{-4} \leq \mathbf{Oh} \leq 4.0 \times 10^{-4}$. For the realization of the non-isothermal behavior the axial wall temperature gradients $\Delta T_w / \Delta Z$ were varied in the range 0.0 - 2.9 K/mm resultung in a variation of the following dimensionless numbers, Table A.11-A.15. So the evaporation number at the contact line was varied as $0.001 \times 10^{-1} \leq \mathbf{E}_{cl} \leq 3.2 \times 10^{-1}$ whereas the evaporation number at the interface as $0 \leq \mathbf{E} \leq 4.32 \times 10^{-2}$. The range of variation of $\mathbf{Ev}_v$ was $0.0 \leq \mathbf{Ev}_v \leq 36.3 \times 10^6$. The ranges of variation of $\mathbf{E}\,\mathbf{Oh}$ and $\mathbf{E}_{cl}\,\mathbf{Oh}$ are $0 \leq \mathbf{E}\,\mathbf{Oh} \leq 9.46 \times 10^{-6}$ and $0.002 \times 10^{-5} \leq \mathbf{E}_{cl}\,\mathbf{Oh} \leq 10.8 \times 10^{-5}$, respectively.

The above given ranges express the essential feature of the experimental investigation: a significant variation of the evaporative dimensionless numbers by more than one magnitude for each liquid whereas the capillary dimensionless numbers were varied insignificantly. The $\mathbf{Oh}$ number was varied by a factor of 1.08 compared to 14.75 of $\mathbf{E}_{cl}$. The factors of the variation with $\mathbf{LCH_4}$ were 1.13 of $\mathbf{Oh}$, 16.71 of $\mathbf{E}_{cl}$, respectively. The corresponding values with $\mathbf{LNe}$ were 1.03 of $\mathbf{Oh}$ and 59.67 of $\mathbf{E}_{cl}$. The $\mathbf{Oh}$ number was varied with $\mathbf{LH_2}$ by a factor of 1.05. Due to the really isothermal experiment experiment with $\mathbf{LH_2}$ with the factor of the variation of $\mathbf{E}_{cl}$ appears untypically big - 4571.

The same applies to the factor of the variation of velocity scale of the capillary-driven flow $U = u_{puR}$ which was used for the non-dimensionlization of the field equations and boundary conditions in Section 3.3.2: $\mathbf{LAr}$ - 1.02, $\mathbf{LCH_4}$ - 1.03, $\mathbf{LNe}$ - 1.003, $\mathbf{LH_2}$ - 1.02, Table A.6-A.10. So this velocity scale is practically constant $U \approx const$, indicating that variation of the interface motion should be attributed to the influence on the velocity by the varied mass flux in the boundary conditions at the the the interface and the contact line.

The present investigation concentrates on the dynamic behavior of the liquid-vapor interface during the global time domain of the reorientation process as defined by [50]. The cause for this restriction was the technical impossibility for the direct investigation of the initial time domain of reorientation defined by [50].

Regarding the isothermal behavior (extrapolation to zero mass flux) one specific goal of the present investigation is to examine whether the oscillation frequency of the cryogenic liquids complies with the presented findings on storable liquids at isothermal conditions. Therefor it is tested whether the frequency lies within the theoretical values $\Omega_n = 3.0$ with the free contact line and $\Omega_n = 3.5$ with the fixed contact line according to Eq.(2.42-2.43) for $\gamma_s \to 0°$ in Chapter (5).

As already stated the steady-state contact angle with evaporation must be larger than the static contact angle [2]. Therefore a comparison theory and experiment is done in Chapter (5) to identify or verify the value of the mobility exponent $m = 3$ in the relation that gives this steady-state contact angle, which I assume as the contact line boundary condition for the field equations in Section (3.3). In the steady-state case the relation has the form given by Eq. (2.23). Actually such experimental tests were suggested by [2] regarding the original contact line relation Eq. (2.20) proposed by them from which Eq. (2.23) is derived. It was shown by [25] that $m = 3$ gives an excellent agreement between theory and experiment in the case of the isothermal droplet spreading. Experiments were performed later by [24] also for the isothermal droplet spreading and values in the narrow range $2.64 \leq m \leq 2.95$ with an average value $m_{av} = 2.8$ were determined. Additional experiments by [24] suggest a value $m = 2.8$ for the non-isothermal droplet spreading due to thermocapillarity.

However, the interface reorientation can be considered not only from the viewpoint of fluid dynamics but also from the viewpoint of system dynamics. Within the system dynamics the interface reorientation can be treated as a step response of a 2nd order linear system, as presented in Chapter 3. This allows to take into account the fundamental relation between the steady-state deflection of the step response and the natural frequency of the oscillations [58], a relation not considered by [50] and [29] by the investigation of the isothermal reorientation. re of the According to this relation given by Eq. (3.85) the square of the frequency is inversely proportional to the stead-state deflection. In the presence of the wall superheat the steady state-value of the interface deflection from its initial equilibrium value would vary corresponding to a steady-state value of the non-isothermal contact angle, the latter one being a function of the wall superheat.

The variation of the contact line boundary condition is expected to lead first to a change of the contact line behavior and a variation of the characteristics of this motion for the same liquid. Alongside with this a variation of the characteristics of the interface center point should also be observed, due to the the coupling of the motion of the two characteristic points pointed out by [50] and [29]. This contrasts to the isothermal reorientation where such variations are only possible through a variation of the test liquids (through the variation of their material properties). The changed contact line behavior demands a specific quantitative characterization of the contact line motion, at least in terms of velocity, after the moment of its maximal deflection which has not been done for the investigation of the isothermal reorientation as by [50] and [29].

The investigation of the contact line motion by [50] was restricted to the velocity of the initial transient rise and to the subsequent time of settling of the oscillatory motion of the contact line. Quantitatively regarded an observation was made of the number of extremal points until the settling of the contact line. However, no investigation of the velocity of this oscillatory motion was performed. It was suggested by [50] that an investigation of the possible variation of initial rise velocity and maximum deflection of the contact line due to the evaporation effects. So these two characteristics alongside with the velocity of the oscillatory motion are investigated, Chapter 5.

Generally speaking, according to the examples of Section 2.5, the action of evaporation on the flow and on the reorientation characteristics would be expected to be opposite to this of capillarity - e.g a decrease of the interface deflection and an increase of the oscillation frequency, as shown by the experimental results in [80] for the steady evaporating meniscus and by the numerical results in [64] for the transient capillary rise with evaporation. The same opposite action is observed by droplet spreading with evaporation as well - variation of the motion of the contact line and the center point with the variation of the wall superheat, increase of the steady-state contact angle. Accordingly an analogous variation of the reorientation characteristics observed under isothermal conditions (due to capillarity) are to be expected due to evaporation on the interface and the contact line [50]. This variation is theoretically treated and experimentally investigated in this work.

3 Theory

3.1 Model assumptions and implications

My assumption of the non-isothermal reorientation as a capillary-driven flow with phase change has implications concerning the formulation of the field equations and the boundary conditions alongside their scaling approach. Interfacial phase change is the only thermal mechanism influencing the liquid flow, through the mass flux terms in the boundary conditions. Thermo-capillary convection and vapor recoil are not included. A boundary condition at the contact line is included accounting for the interfacial phase change due to the local wall superheat [2]. The velocity scale by [50], treating the isothermal reorientation as capillary-driven, is chosen for the liquid flow.

The transient and the steady-state characteristics of the reorientation from the experiments would be dependent on the mass flux (and parameter) variation in the boundary conditions in dimensional form or on the variation of the dimensionless numbers arising from their scaling. The mathematical form of these conditions is changed in the presence of interfacial phase change. Additional terms depending on the interfacial mass flux is included in the mass balance boundary condition at the interface and the contact line. This in turn would lead not just to a variation of the value of the velocity field, which is the solution of the Navier-Stokes equation in the integration region, but to a change of its spatial distribution and temporal evolution. Accordingly a change of the flow pattern as a whole would occur, manifested as a change of the interface motion too. The additional terms accounting for the non-isothermal conditions vanish for a zero mass flux. So it is ensured that the field equations and the boundary conditions presented in this chapter describe the flow pattern and the interface motion characteristic for perfectly wetting liquids at isothermal conditions as well, precisely as the results by [50] and [29].

Since I make a particular selection of the contact line boundary condition based on [2], this would imply that the value of the fourth power of the contact angle describing the final steady-state interface would be proportional to the mass flux at the contact line or to the the corresponding dimensionless number.

From the viewpoint of the system dynamics I approximate the reorientation process with a linear step response of second order based on [58] since it describes explicitly the process of transition from one steady-state to another after a step change, as in the case of the reorientation. I extend so the substitution models of reorientation used by [50] and [29]. The implication would be that the square of the oscillation frequency of the interface is inversely proportional to the interface deflection at the final steady-state.

For the formulation of the field equations and the corresponding boundary conditions specific assumptions are made. The flow is non-isotherm, however the temperature dependent variation material properties of the liquid, vapor and the solid phases is disregarded during the reorientation (**A1**). The flow is assumed to be axis-symmetric (**A2**). The transition from normal gravity to compensated gravity is considered to be as a jump by a step reduction of the value of the BOND number (**A3**). Both fluid phases are assumed to be Newtonian (**A4**). The liquid phase is considered incompressible (**A4**). The vapor phase considered as an ideal gas and therefore in principle compressible, however, with the additional assumption of weak compressibility (**A5**). Both fluid phases are assumed to comprise a single species (pure liquid) system (**A6**). The liquid is assumed to be perfectly wetting with a zero static contact angle (**A7**). There is mass transfer across the liquid-vapor interface due to phase change effects of evaporation and condensation (**A8**). The temperature along the interface is equal to the saturation temperature (**A9**). Interfacial phase change is the only thermal mechanism included affecting the flow (**A10**). There is no mass transfer across the liquid-solid and the vapor-solid interface (**A11**). The cylinder wall is considered smooth and rigid (**A12**). There is heat transfer between the cylinder wall and the fluid phases (**A13**). The liquid-vapor interface is considered as a two-dimensional surface with no structure or thickness (**A14**).

3.2 Interface configurations during the non-isothermal reorientation

The process of the reorientation in a partially filled cylinder is the transition from the initial steady-state of the interface under 1g to a steady-state under microgravity, depicted in Fig. (3.1). So between the two steady-states there is a transient state with an oscillating interface characterized by the coordinate of the the center point $Z_c(t)$ and the coordinate of the contact line $Z_w(t)$. The derivative of the contact line coordinate is the contact line velocity $u_{cl} = Z_w(t)/dt$. These three variables are investigated in the present work. The corresponding angle between the interface and the wall at the position of the contact line is the dynamic contact angle γ_d. An axis-symmetric coordinate system is originating at the position of the center point of the 1g interface. There is no analytical approach describing the transient interface, in contrast to other cases such as the thin droplet spreading.

The other goal of the present section is to present the initial 1g and the final μg steady-state of the interface during the reorientation. The mathematical representation here is based on the formalisms describing the isothermal case where only the action of gravity and capillarity is considered. The interfaces under such conditions are static interfaces, the steady state of the interface corresponds to a static equilibrium. Under the action of interfacial evaporation a steady-state interface would require a flow driven by the evaporation so that the steady-state of the interface

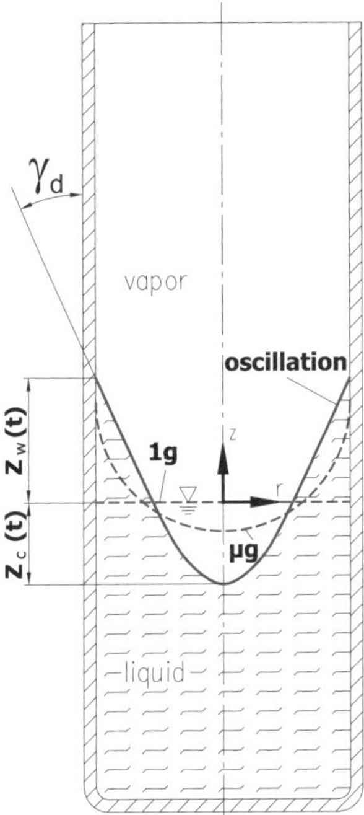

Fig. 3.1. Sketch of a partially filled cylinder for the investigation of the reorientation with the steady-state interfaces by 1g and μg, and the transient oscillating interface.

would correspond to a dynamic equilibrium of the balance between by capillarity (spreading) and evaporation (receding) [2], [64] and [54].

Since no analytical results are known to me in the presence of evaporation I nevertheless assume that the equations for the isothermal case are still a useful approximation of the non-isothermal reorientation. First, the initial wall superheat is so small that the influence of the evaporation on the initial interface, the contact angle and therewith the contact line position, can be regarded as negligible. Second, the final interface can be assumed to be still spherical defined by the steady-state value γ_{ss} of the contact angle, corresponding to the balance of capillarity and evaporation [2] as mentioned above, substituted for static contact angle γ_s in the equations. The other characteristic quantities are the final steady-state deflections of the center point Z_{css} and the contact line Z_{wss}, respectively.

According to assumption (**A3**) a step reduction of the BOND number occurs by the transition from terrestrial to compensated gravity. The BOND number is defined in a cylindrical geometry as

$$\mathbf{Bo} = \frac{\rho\, k_{zi}\, R^2}{\sigma} \tag{3.1}$$

where ρ, σ, k_{zi} and R are the liquid density, surface tension, the acceleration due to gravity and the cylinder radius, respectively. **Bo** can be regarded as the ratio between the hydrostatic and the capillary forces.

I will explain the relation between the value of the BOND number and the steady-state interfaces in the isothermal case. The initial interface corresponds to a high value, **Bo** $\gg 1$. This is is the result of the dominance of the hydrostatic forces. The capillary forces are negligible except at the wall where the interface is curved intersecting the wall with an initial capillary rise Z_{w0} and with a initial steady-state contact angle, the static contact angle γ_s. Away from the wall the interface is flat. The final interface corresponds to a low value, **Bo** $\ll 1$. It is the result of the dominance of the capillary forces where the hydrostatic forces are negligible. The interface is spherical with a constant curvature with a steady-state contact angle at the wall equal to the static contact angle γ_s, too.

3.2.1 Steady-state interface under terrestrial gravity

The mean curvature $2k$ of the interface in the cylindrical coordinate system of the experimental setup can be expressed as

$$2k = \frac{1}{r}\frac{d}{dr}\left[\frac{r\dfrac{dh}{dr}}{\left[1+\left(\dfrac{dh}{dr}\right)^2\right]^{1/2}}\right] \tag{3.2}$$

where $h(r)$ is the function of the interface [50]. Further it can be shown that $\sigma 2H = C + \rho k_{zi}h(r)$ where k_{zi} is applied acceleration and $\rho = \rho - \rho_v$ the density difference between the liquid and vapor.

Non-dimensionalization with $r^* = r/R$, $h^* = h/R$, $\Psi = CR/\sigma$ and the Bond number **Bo** leads to the non-dimensional equation for the initial interface as well as any other static interface under k_{zi} acceleration:

$$\frac{1}{r^*}\frac{\dfrac{dh^*}{dr^*}}{\left[1+\left(\dfrac{dh^*}{dr^*}\right)^2\right]^{1/2}} + \frac{\dfrac{d^2h^*}{dr^{*2}}}{\left[1+\left(\dfrac{dh^*}{dr^*}\right)^2\right]^{3/2}} = \Psi + \mathbf{Bo}\,h^* \tag{3.3}$$

While no analytical solution can be found for Eq. 3.3, numerical solutions can be obtained for the boundary condition at the wall

$$\left.\frac{dh^*}{dr^*}\right|_{r^*=1} = \cot\gamma_s \tag{3.4}$$

given by the static contact angle γ_s. In this manner the static interface in cylindrical containers is only a function of **Bo** and γ_s as a solution of the problem defined by Eq.(3.3) an Eq. (3.4).

At the wall the capillary and the hydrostatic forces become balanced ($\mathbf{Bo} = 1$) resulting in a initial rise of the contact line Z_{w0} of the order of

$$Z_{w0} \propto L_c \sqrt{2(1 - \sin \gamma_s)} \tag{3.5}$$

where $L_c = \sqrt{\sigma/(\rho k_{zi})}$ is the capillary length scale of the system [48]. Concerning the reorientation, the influence of the evaporation on the initial interface can be regarded as negligible. The initial wall superheat is small enough that its influence both on the contact angle (Eq.(2.23)) as well as on the initial rise can be ignored so that Eq.(3.3), Eq. (3.4) and Eq. 3.5 approximate very good the initial interface of the non-isothermal reorientation.

3.2.2 Steady-state interface under microgravity

The step reduction of gravity results in a step reduction of the Bond number to a value $\mathbf{Bo} \ll 1$. Under isothermal conditions this interface, which is the final steady-state, is spherical [82] with a dimensionless radius β which is function only of the static contact angle γ_s. I assume that the final interface of the non-isothermal reorientation is still spherical based on the literature. A constant curvature was assumed by [74] for the macroscopic portion of the evaporating meniscus in grooves. Regarding the dynamics of an evaporating meniscus in small capillaries due to the evaporation-driven flow investigated by [65] the interface was assumed as nearly spherical with a certain radius. I assume therefore that the steady-state value γ_{ss} of the contact angle [2], arising from the balance of capillarity and evaporation, can be substituted instead of γ_s in the following equations.

Accordingly, the relation for the mentioned dimensionless radius β

$$\beta = \frac{1}{\cos \gamma_{ss}} \tag{3.6}$$

depending only the static contact angle γ_{ss}.

The dimensionless difference between the positions of the contact line $h^+ (r^* = 1, t \to \infty) = Z_{wss}/R$ and the center point $h^+ (r^* = 0, t \to \infty) = Z_{css}/R$ reads

$$\alpha = h^+ (r^* = 1, t \to \infty) - h^+ (r^* = 0, t \to \infty) = \frac{1 - \sin \gamma_{ss}}{\cos \gamma_{ss}} \tag{3.7}$$

by which the dimensionless radius is given

$$\beta = \frac{1 + \alpha^2}{2\alpha} \tag{3.8}$$

by substitution of γ_{ss} by α. Accordingly the the static contact angle γ_{ss} can be expressed as

$$\gamma_{ss} = \arccos \frac{2\alpha}{1 + \alpha^2} \tag{3.9}$$

which relation is very useful for an evaluation of γ_{ss} from the final interface when the latter is observed experimentally.

An analytical solution of Eq. (3.3) given by [4] for $\mathbf{Bo} = 0$ describing the interface as a function of the non-dimensionalized container radius r^* and the static contact angle γ_s. The solution reads in dimensionless terms substituting γ_s with γ_{ss}

$$h^+(r^*, t \to \infty) = \frac{2(1 - \sin^3 \gamma_{ss})}{3 \cos^3 \gamma_{ss}} - \frac{1}{\cos \gamma_{ss}} \left[1 - (r^* \cos \gamma_{ss})^2 \right]^{1/2} \qquad (3.10)$$

where $h^+ = h(r)/R$ is the non-dimensionalized height function of the final interface. Accordingly the final position of the center point is given by

$$h^+(r^* = 0, t \to \infty) = \frac{Z_{css}}{R} = -\frac{1}{\cos \gamma_{ss}} \left[1 - \frac{2(1 - \sin^3 \gamma_{ss})}{3 \cos^2 \gamma_{ss}} \right] \qquad (3.11)$$

where $Z_{c,ss}$ is the final position of the center point in dimensional terms. The final position of the contact line reads

$$h^+(r^* = 1, t \to \infty) = \frac{Z_{wss}}{R} = \frac{2 - 3 \sin \gamma_{ss} + \sin^3 \gamma_{ss}}{3 \cos^3 \gamma_{ss}} \qquad (3.12)$$

with $Z_{c,ss}$ the dimensional final position of the contact line, respectively.

3.3 Field equations, boundary conditions and scaling

3.3.1 Dimensional form of the field equations and boundary conditions

I propose a dimensional formulation of the field equations describing the reorientation in an incompressible form for the liquid phase and in a weak compressible from for the vapor phase, based on the scale analysis for small Mach numbers presented in [43]. Interfacial evaporation/condensation is the only thermal mechanism influencing the liquid flow. At the contact line I impose a boundary condition based on Eq. (2.20), through which the influence of the interfacial mass flux due to the wall superheat on the flow is taken explicitly into account. This the essential extension of the mathematical formalism I presented in [45].

Therefore the formalism below represents a full equation system describing the non-isothermal reorientation, which is an extension of formalisms used by [83], [50] and [29] for the description of the isothermal reorientation.

The governing equations representing the physical situation for the liquid phase are the Navier-Stokes equations for an incompressible Newtonian liquid are

$$\rho \left(\frac{\partial \mathbf{u}}{\partial t} + \mathbf{u} \cdot \nabla \mathbf{u} \right) - \mu \nabla^2 \mathbf{u} + \nabla P = \mathbf{f} = 0, \qquad (3.13)$$

$$\nabla \cdot \mathbf{u} = 0, \qquad (3.14)$$

$$\rho \, c_p \left(\frac{\partial T}{\partial t} + \mathbf{u} \cdot \nabla T \right) - \lambda \nabla^2 T = 0, \qquad (3.15)$$

where $\mathbf{u}$, P and T are the field functions of the liquid velocity, pressure and temperature. The constant-pressure specific heat and the thermal conductivity of the liquid are c_p and λ, respectively. The external body forces $\mathbf{f}$ are set to 0 at the step reduction of gravity.

At the interface one has the following boundary conditions

$$\mathbf{T}\mathbf{n} = \sigma\, k\, \mathbf{n} \tag{3.16}$$

$$u_I = \mathbf{u} \cdot \mathbf{n} - \frac{1}{\rho} j \tag{3.17}$$

$$T = T_I \tag{3.18}$$

where $\mathbf{T}$, $\mathbf{n}$ and k are the Cauchy stress tensor, the unit vector normal to the interface and the curvature of the interface, defined as $k = 1/R_m$ with R_m determined by Eq. (2.3)[81]. The quantity u_I is the normal component of the interface velocity and j is the mass flux through the interface.

I choose according to [2] a general boundary condition at the contact line

$$u_{cl} = -\frac{j_{cl}}{\rho \sin \gamma_d} + u\,|_{cl} \tag{3.19}$$

where $u\,|_{cl}$ is the liquid velocity at the contact line.

Based on Eq. (2.20) the above equation becomes

$$u_{cl} = -\frac{j_{cl}}{\rho \sin \gamma_d} + \kappa\,(\gamma_d - \gamma_s)^m = -\frac{j_{cl}}{\rho \sin \gamma_d} + \kappa\gamma_d^3 \tag{3.20}$$

assuming that the cryogenic liquids are perfectly wetting with no contact hysteresis ($\gamma_s = \gamma_{sa} = \gamma_{sr} = 0$). Additionally I assume that $m = 3$. This is the specific boundary condition according to the hypothesis by [2] of $u\,|_{cl}$ and my assumptions.

The mass flux at the contact line j_{cl} is calculated according to Eq. (2.18) and proportional to the wall superheat $T_{w,cl} - T_S$. So the influence of the local wall superheat on the flow is given by the above Eq. (3.19) and Eq.(3.20) explicitly. These relations provide for all contact line motions - isothermal/non-isothermal and stationary/transient. Through Eq. (3.20) the dynamic contact angle is determined unambiguously in relation to the contact line velocity u_{cl} and the contact line mass flux j_{cl} (or $T_{w,cl} - T_S$) given that κ and m are known.

I just to note here that one can substitute the linear temperature distribution in the wall from Eq. (4.17) in the above equation

$$u_{cl} = -\frac{K}{\rho \sin \gamma_d}\left(T_{WIF} - T_S + \frac{\Delta T_w}{\Delta Z}Z_w\right) + \kappa\gamma_d^3 \tag{3.21}$$

where Z_w is the position of the contact line and T_{WIF} is the measured temperature on the outside cylinder wall at the liquid fill level shown in Figure 4.4 (T_{wf} in Figure

4.3). The above equation relates explicitly u_{cl} and γ_d to the axial wall gradients $\Delta T_w/\Delta Z$. So it shows the explicit influence on the liquid flow of the parameter $\Delta T_w/\Delta Z$ varied in the experiments.

As a result of the scale analysis presented in [43], I assume the flow in the vapor phase to be characterized by the weakly compressible counterparts of the governing equations for the liquid phase,

$$\rho_v\left(\frac{\partial \mathbf{u}_v}{\partial t}+\mathbf{u}_v\cdot\nabla\mathbf{u}_v\right)-\mu_v\nabla^2\mathbf{u}_v+\nabla P_{v,hyd}=\mathbf{f}_v=0\,,\tag{3.22}$$

$$\nabla\cdot\mathbf{u}_v=-c\,,\tag{3.23}$$

$$\rho_v c_{p,v}\left(\frac{\partial T_v}{\partial t}+\mathbf{u}_v\cdot\nabla T_v\right)-\lambda_v\nabla^2 T_v=\frac{\partial P_v}{\partial t}\,,\tag{3.24}$$

where $\mathbf{u}_v$ and T_v are the field functions of the vapor velocity and temperature. The external body forces acting the vapor $\mathbf{f}_v$ are also set to zero at the step reduction of gravity as by the liquid. P_v is the vapor pressure and is approximated by the ideal gas law, Eq. (3.31), as discussed later. P_v is the quantity measured in my experiments. $P_{v,hyd}$ is a pressure quantity in a hydrodynamical sense meaning it is relevant only to the flows in the vapor, [43]. The density and dynamic viscosity of the vapor are ρ_v and μ_v, respectively.

The evolution of the thermodynamic pressure in vapor P_v is governed by the ordinary differential equation

$$\frac{1}{P_v}\frac{\partial P_v}{\partial t}=\xi c\tag{3.25}$$

with ξ being the adiabatic exponent $\xi=c_{p,v}/c_{v,v}$ where $c_{v,v}$ and $c_{p,v}$ is the constant-volume specific heat and constant-pressure specific heat of the vapor, respectively.

By c I denote the spatially constant compression rate of the vapor which is due to the interface movement and the mass flux j in a closed system,

$$c=\frac{1}{V_v}\int_{\Gamma_I}\left(\mathbf{u}\cdot\mathbf{n}+\frac{\rho-\rho_v}{\rho\rho_v}j\right)d\Gamma_I\,,\tag{3.26}$$

where V_v is the volume occupied by the vapor, Γ_I is the surface area of the interface and the thermal conductivity of the vapor is λ_v.

The boundary conditions on the interface from the vapor phase read

$$\mathbf{u}_v=\mathbf{u}-\frac{\rho-\rho_v}{\rho\rho_v}j\mathbf{n}\,,\tag{3.27}$$

$$T_v=T_I\,,\tag{3.28}$$

Again I assume the temperature to correspond to T_I along the interface.

The mass flux across the interface (with the exception of the contact line as above mentioned) is modeled according Eq. (2.19) through the jump in the conductive heat flux across the interface. The respective equation is repeated here

$$j = -\frac{1}{\Delta h_{lv}} \left[\lambda \, \partial_{\mathbf{n}} T + \lambda_v \, \partial_{\mathbf{n}} T_v \right], \tag{3.29}$$

where $\partial_{\mathbf{n}}$ denotes the normal derivative from the liquid to the vapor. The mass flux at the contact line is modeled through Eq. (2.18) repeated here

$$j_{cl} = K^{-1}(T_{w,cl} - T_S) = \left(\frac{\alpha \, \rho_v \, \Delta h_{lv}}{T_S^{3/2}} \right) \left(\frac{M_m}{2 \pi R_u} \right)^{\frac{1}{2}} (T_{w,cl} - T_S) \tag{3.30}$$

I consider that the ideal gas law approximates the thermodynamic behavior of the vapor so that the equation of state in the vapor holds

$$P_v = \rho_v \, R_s T_v, \tag{3.31}$$

where R_s is the specific gas constant. I assume the saturation temperature T_S to be related to the pressure of the vapor phase P_v through the Clausius-Clapeyron equation.

$$\frac{dP_v}{dT} = \frac{\Delta h_v \, \rho \, \rho_v}{T_S \, (\rho - \rho_v)}. \tag{3.32}$$

At last I assume that in the solid phase of my problem, the wall, the energy equation reads

$$\rho_w \, c_{p,w} \frac{\partial T_w}{\partial t} - \lambda_w \, \nabla^2 T_w = 0, \tag{3.33}$$

where T_w is the wall temperature. The density, the constant-pressure specific heat and the thermal conductivity of the wall are ρ_w, $c_{p,w}$ and λ_w, respectively.

3.3.2 Non-dimensional form of the field equations and boundary conditions

In the previous section the dimensional field equations and boundary conditions were presented in their general form following [45] and [2]. Here I present the field equations and their boundary conditions at the interface in a non-dimensional form following and adapting [44] and [3] to the case two-phase flow of pure liquids. Simultaneously I took into consideration the scaling approaches by [50] [29], [11] and [2].

As already mentioned I consider the non-isothermal reorientation as a capillary-driven flow with interfacial phase change. By the scaling of the dimensional equations only in the global time domain with the cylinder radius R as a length scale [50]. Accordingly I assume that the velocity scale of the liquid flow is given by

$$U = u_{puR} = \sqrt{\frac{\sigma}{\rho R}} \tag{3.34}$$

which is the characteristic velocity used by [50] for the mathematical treatment of the isothermal reorientation as a capillary-driven flow problem. The time scale is

$\tau = R/U = \sqrt{\rho R^3/\sigma}$. As a length scale I choose R, since I am interested in the global time domain of the reorientation.

I assume two length scales: the cylinder radius R and the thickness of the interfacial thermal boundary layer δ_T of the quiescent interface. The thickness can be estimated through $\delta_T \approx \sqrt{(\lambda/\rho\,c_p)t_T}$ by [15]. The quantity t_T is the time needed for the buildup of the thermal boundary layer, which by the reorientation experiments is equal to the time span from the start of the heating to the moment of the step reduction of gravity, Table A.1-A.5.

The WEBER number **We** and both OHNESORGE numbers, **Oh** in liquid and **Oh$_v$** in vapor, as well as the PECLET number in the wall **Pe$_w$** are calculated through R, Table (3.1). However, the WEBER number is set to unity **We** $= 1$ due to the choice of the velocity scale U.

The mass flux scale J can be defined through the balance of the heat conducted through the interfacial thermal boundary layer (from the undercooled isothermal liquid bulk to the interface) and the heat of evaporization at the interface analogous to [11] and [2]

$$J = \frac{\lambda\Theta}{\Delta\,h_{lv}\delta_T} \tag{3.35}$$

where δ_T is used as a length scale for the calculation of J instead of the thickness of the liquid layer over the solid in [11] or of the thickness of liquid drop on the solid surface [2]. The EVAPORATION heat flux number in the liquid **Ev** is calculated with δ_T, whereas EVAPORATION heat flux number in the vapor **Ev$_v$** is calculated with R, Table 3.1. The EVAPORATION heat flux number is set to unity **Ev** $= 1$ due to the choice of the mass flux scale J. The mass flux scale at the contact line

$$J_{cl} = \frac{\lambda\Theta_w}{\Delta\,h_{lv}\delta_T} \tag{3.36}$$

is calculated through the characteristic wall superheat Θ_w.

The non-dimensionalization of all equations and boundary conditions demands additionally these scales for the following physical physical quantities. Reference densities for the liquid and vapor are chosen, ρ_0 and $\rho_{0,v}$, respectively. Scales of the difference between the temperature of the liquid, vapor and the wall and the saturation temperature T_S are defined as $\Theta = T_S - T_{ch}$, $\Theta_v = T_{v,ch} - T_S$ and $\Theta_w = T_{w,ch} - T_S$, respectively [11], [2]. The characteristic wall superheat is given also by Θ_w. The quantities T_{ch}, $T_{v,ch}$ and $T_{w,ch}$ are characteristic temperatures of liquid, vapor and wall, respectively. Considering the experiments I assume that the $T_{ch} = T_L$ with geometry A and $T_{ch} \approx T_{WL1}$ with geometry B. As a characteristic temperature of the vapor I consider $T_{v,ch} = T_V$ with geometry A and $T_{v,ch} = T_{V4}$ with geometry B. The characteristic temperature of the wall I assume to be $T_{w,ch} = T_w(Z = R)$.

Several scales of the pressure are required. The pressure scale in the liquid is chosen as $\rho_0 U^2$. In the vapor there is a hydrodynamic pressure scale in the vapor $\rho_{0,v} U^2$ and a thermodynamic pressure scale $\rho_{0,v} R_s\,\Theta_v$.

In this manner the field quantities, the differential operators and the curvature can be non-dimensionalized through the above given scales as

$$\mathbf{u} = \hat{\mathbf{u}}\,U,\ P = \hat{P}\rho_0 U^2,\ T - T_S = \hat{T}\,\Theta,\ j = \hat{j}\,J,\ P_v = \hat{P}_v\,\rho_{0,v}\,R_s\,T_S,$$

$$t = \hat{t}\,\tau,\ \mathbf{T} = \hat{\mathbf{T}}\,\rho_0 U^2,\ \nabla = \frac{\hat{\nabla}}{L},\ \nabla^2 = \frac{\hat{\nabla}^2}{L^2},\ k = \frac{\hat{k}}{L} \tag{3.37}$$

where the quantities with the hat are the corresponding dimensionless quantities. L is the length scale, which is either $L = R$ or $L = \delta_T$.

Substituting the dimensional quantities from the above relations into the field equations for the liquid gives after little transformation

$$\frac{R}{U\tau}\frac{\partial \hat{\mathbf{u}}}{\partial \hat{t}} + \hat{\mathbf{u}}\cdot\hat{\nabla}\hat{\mathbf{u}} - \frac{\mu}{\rho_0 U R}\hat{\nabla}^2\hat{\mathbf{u}} + \hat{\nabla}\hat{P} = 0, \tag{3.38}$$

$$\hat{\nabla}\cdot\hat{\mathbf{u}} = 0, \tag{3.39}$$

$$\frac{\partial \hat{T}}{\partial \hat{t}} + \hat{\mathbf{u}}\cdot\hat{\nabla}\hat{T} - \frac{\lambda}{\rho_0 c_p U L}\hat{\nabla}^2\hat{T} = 0, \tag{3.40}$$

At the interface Γ_S the boundary conditions become in terms of dimensionless quantities after minimal transformation

$$\hat{\mathbf{T}}\mathbf{n} = \frac{\rho_0 U^2}{R}\,\sigma\,\hat{k}\mathbf{n}, \tag{3.41}$$

$$\hat{u}_I = \hat{\mathbf{u}}\cdot\mathbf{n} - \frac{J}{\rho_0 U}\hat{j}, \tag{3.42}$$

$$\hat{T} = \hat{T}_S(P_v), \tag{3.43}$$

where $\hat{u}_I$ and $\hat{T}_S$ are the dimensionless normal component of the interface velocity and the dimensionless saturation temperature, respectively. The dimensionless saturation temperature $\hat{T}_S$ is related to the dimensionless thermodynamic vapor pressure $\hat{P}_v$ through the Clausius-Clapeyron equation in the form:

$$\hat{T}_S(\hat{P}_v) = \frac{1}{\Theta_v}\left(-\frac{\Delta h_{lv}}{R_{s,v}\log\left(\frac{\hat{P}_v\rho_{0,v}R_s T_0}{b}\right)} - T_0 \right) \tag{3.44}$$

The general boundary condition at the contact line reads

$$\hat{u}_{cl} = -\frac{J_{cl}}{\rho_0 U}\frac{\hat{j}_{cl}}{\sin\gamma_d} + \hat{u}\,|_{cl} \tag{3.45}$$

whreas the specific one reads

$$\hat{u}_{cl} = -\frac{J_{cl}}{\rho_0 U}\frac{\hat{j}_{cl}}{\sin\gamma_d} + \frac{\kappa\varepsilon^3}{U}\hat{\gamma}_d^3 = -\frac{J_{cl}}{\rho_0 U}\frac{\hat{j}_{cl}}{\sin\gamma_d} + \hat{\kappa}\hat{\gamma}_d^3 \tag{3.46}$$

where ε is the aspect ratio of the interface at the contact line and the dynamic contact scales with it like $\gamma_d = \varepsilon\,\hat{\gamma}_d$ [2]. The quantity $\hat{\kappa}$ is the non-dimensionalized κ, analogous to Eq. (2.22). A possible definition for the aspect ratio by the reorientation would be $\varepsilon = \delta_T/R$ based on [2].

The field equations in the vapor phase become through the dimensionless quantities in Eq. (3.37)

$$\frac{R}{U\tau}\frac{\partial \hat{\mathbf{u}}_v}{\partial \hat{t}} + \hat{\mathbf{u}}_v \cdot \hat{\nabla}\hat{\mathbf{u}}_v - \frac{\mu}{\rho_{0,v}UR}\hat{\nabla}^2\hat{\mathbf{u}}_v + \hat{\nabla}\hat{P}_{v,hyd} = 0\,, \tag{3.47}$$

$$\hat{\nabla}\cdot\hat{\mathbf{u}}_v = -\hat{c}\,, \tag{3.48}$$

$$\frac{\partial \hat{T}_v}{\partial \hat{t}} + \hat{\mathbf{u}}_v \cdot \hat{\nabla}\hat{T}_v - \frac{\lambda}{\rho_{0,v}c_{p,v}UR}\hat{\nabla}^2\hat{T}_v = \frac{\rho_{0,v}R_s}{c_{p,v}\Theta_v}\frac{\partial \hat{P}_v}{\partial \hat{t}}\,, \tag{3.49}$$

Accordingly the evolution of the dimensionless thermodynamic pressure in vapor $\hat{P}_v$ would be governed by the ordinary differential equation

$$\frac{1}{\hat{P}_v}\frac{\partial \hat{P}_v}{\partial \hat{t}} = \xi\,\hat{c} \tag{3.50}$$

with $\hat{c}$ being the spatially constant dimensionless compression rate. The relation for $\hat{c}$ reads

$$\hat{c} = \frac{1}{V_v}\int_{\Gamma_I}\hat{\mathbf{u}}_l\cdot\mathbf{n} + \left(\frac{\rho_0}{\rho_{0,v}}-1\right)\frac{J}{\rho_0 U}\,\mathrm{d}\Gamma_I\,, \tag{3.51}$$

The boundary conditions for the vapor flow at the interface become in dimensionless terms

$$\hat{\mathbf{u}}_v = \hat{\mathbf{u}}_l + \left(\frac{\rho_0}{\rho_{0,v}}-1\right)\frac{J}{\rho_0 U}\hat{j}\mathbf{n}\,, \tag{3.52}$$

$$\hat{T} = \hat{T}_S(\hat{P}_v)\,, \tag{3.53}$$

The dimensionless evaporative mass flux across the interface $\hat{j}$ is calculated through

$$\hat{j} = -\frac{\lambda\,\Theta}{\Delta\,h_{lv}J\,\delta_T}\partial_{\mathbf{n}}\hat{T} + \frac{\lambda_v\Theta_v}{\Delta\,h_{lv}JR}\partial_{\mathbf{n}}\hat{T}_v\,, \tag{3.54}$$

using that the liquid and the vapor temperatures scale like $T - T_S = \hat{T}\,\Theta$ and $T_v - T_S = \hat{T}_v\Theta_v$. Accordingly different length scales were chosen for the non-dimensionalization of the differential operator in the liquid and in the vapor, δ_T and R, respectively.

The dimensionless evaporative flux at the contact line is

$$\hat{j}_{cl} = K^{-1}\frac{\Delta\,h_{lv}\,\delta_T}{\lambda}\hat{T}_w = \hat{K}^{-1}\hat{T}_w \tag{3.55}$$

since the contact line mass flux and the local wall superheat scale like $j_{cl} = \hat{j}_{cl}\lambda\,\Theta_w/(\Delta\,h_{lv}\,\delta_T)$ and $T_{w,cl} - T_S = \Theta_w\hat{T}_w$, respectively [11], [2]. The dimensionless non-equilibrium parameter is $\hat{K}$.

The energy equation in the wall becomes in dimensionless terms

$$\frac{\partial \hat{T}_w}{\partial \hat{t}} - \frac{\lambda_w}{\rho_w \, c_{p,w} \, U \, R} \hat{\nabla}^2 \hat{T}_w = 0 \,, \tag{3.56}$$

Through the above scales the following dimensionless numbers arise, summarized in Table (3.1). In this manner the factors in the above dimensionless equations presented as combinations of these dimensionless numbers.

Thus in terms of the above defined dimensionless numbers the field equations in the liquid phase read

$$\frac{\partial \mathbf{u}}{\partial \hat{t}} + \hat{\mathbf{u}} \cdot \hat{\nabla}^2 \hat{\mathbf{u}} - \mathbf{Oh} \, \hat{\nabla}^2 \hat{\mathbf{u}} + \hat{\nabla} \hat{P} = 0 \,, \tag{3.57}$$

$$\hat{\nabla} \cdot \hat{\mathbf{u}} = 0 \,, \tag{3.58}$$

$$\frac{\partial \hat{T}}{\partial \hat{t}} + \hat{\mathbf{u}} \cdot \hat{\nabla} \hat{T} - \frac{\mathbf{Oh}}{\mathbf{Pr}} \hat{\nabla}^2 \hat{T} = 0 \,, \tag{3.59}$$

In terms of dimensionless numbers the boundary conditions become

$$\hat{\mathbf{T}}\mathbf{n} = \frac{1}{\mathbf{We}} \hat{k} \mathbf{n} = \hat{k} \mathbf{n} \,, \tag{3.60}$$

$$\tag{3.61}$$

$$\hat{u}_I = \hat{\mathbf{u}} \cdot \mathbf{n} - \mathbf{E} \, \mathbf{Oh} \, \hat{j} \,, \tag{3.62}$$

$$\tag{3.63}$$

$$\hat{T} = \hat{T}_S(P_v) \,, \tag{3.64}$$

taking into account $\mathbf{We} = 1$.

The general boundary condition at the contact line becomes in dimensionless numbers

$$\hat{u}_{cl} = -\mathbf{E}_{cl} \, \mathbf{Oh} \, \frac{\hat{j}_{cl}}{\sin \gamma_d} + \hat{u} \, |_{cl} \tag{3.65}$$

whereas the specific contact line boundary condition reads

$$\hat{u}_{cl} = -\mathbf{E}_{cl} \, \mathbf{Oh} \, \frac{\hat{j}_{cl}}{\sin \gamma_d} + \hat{\kappa} \hat{\gamma}_d^3 \tag{3.66}$$

Both Eq. (3.65) and by Eq. (3.66) are of great importance for the current investigation. These are the only relations, here in a non-dimensionalized form, with an explicit dependence on the wall superheat through the corresponding dimensionless number $\mathbf{E}_{cl}$.

Here I want to make a remark on the meaning of $\mathbf{E}_{cl} \mathbf{Oh}$. It can be rewritten as

$$\mathbf{E}_{cl} \, \mathbf{Oh} = \frac{J_{cl}/\rho_0}{U} = \frac{U_{ev,cl}}{U} \tag{3.67}$$

Table 3.1. Dimensionless numbers characterizing the reorientation. The definitions of the temperature scales in the liquid, vapor and solid, used for the calculation of the corresponding numbers, are given for both geometries.

$$\mathbf{Ev} = \frac{\Delta h_{lv} J \, \delta_T}{\lambda \, \Theta} = 1 \qquad\qquad\qquad \text{EVAPORATION heat flux number}$$

$$\mathbf{Ev_v} = \frac{\Delta h_{lv} J R}{\lambda_v \, \Theta_v} \qquad\qquad\qquad \text{EVAPORATION heat flux number in vapor}$$

$$\mathbf{E} = \frac{\lambda \, \Theta}{\Delta h_{lv} \, \mu} \qquad\qquad\qquad \text{EVAPORATION number}$$

$$\mathbf{E_{cl}} = \frac{\lambda \, \Theta_w}{\Delta \, h_{lv} \, \mu} \qquad\qquad\qquad \text{EVAPORATION number at the contact line}$$

$$\mathbf{Oh} = \frac{\mu}{\rho_0 U R} \qquad\qquad\qquad \text{OHNESORGE number}$$

$$\mathbf{Oh_v} = \frac{\mu_v}{\rho_{0,v} U R} \qquad\qquad\qquad \text{OHNESORGE number in vapor}$$

$$\mathbf{Pe_w} = \frac{\lambda_w}{\rho_{0,w} \, c_{p,w} \, U R} \qquad\qquad\qquad \text{PECLET number in wall}$$

$$\mathbf{Pr} = \frac{\mu \, c_p}{\lambda} \qquad\qquad\qquad \text{PRANDTL number}$$

$$\mathbf{Pr_v} = \frac{\mu_v \, c_{p,v}}{\lambda_v} \qquad\qquad\qquad \text{PRANDTL number in vapor}$$

$$\mathbf{Ptd_v} = \frac{c_{p,v} \Theta_v}{R_s T_S} = \frac{\gamma \Theta_v}{(\gamma - 1) T_S} \qquad\qquad\qquad \text{Thermodynamic PRESSURE number in vapor}$$

$$\mathbf{R_\rho} = \frac{\rho_0}{\rho_{0,v}} \qquad\qquad\qquad \text{DENSITY ratio number}$$

$$\mathbf{We} = \frac{\rho_0 U^2 R}{\sigma} = 1 \qquad\qquad\qquad \text{WEBER number}$$

$$\Theta = T_S - T_L \qquad\qquad\qquad \text{Temperature scale in the liquid with geometry A}$$

$$\Theta = T_S - T_{WL1} \qquad\qquad\qquad \text{Temperature scale in the liquid with geometry B}$$

$$\Theta_v = T_V - T_S \qquad\qquad\qquad \text{Temperature scale in the vapor with geometry A}$$

$$\Theta_v = T_{V4} - T_S \qquad\qquad\qquad \text{Temperature scale in the vapor with geometry B}$$

$$\Theta_w = T_w(Z = R) - T_S = T_{WIF} + \tfrac{\Delta T_w}{\Delta Z} R - T_S \quad \text{Temperature scale in the solid with both geometries}$$

which is in principle identical with $\hat{C}^*$ in Eq. (2.34)[13] and proportional to the inverse of C^* in Eq. (2.32)[2]. Thereby $J_{cl}/\rho_0 = U_{cl,ev}$ is interpreted as the evaporative velocity scale $U_{cl,ev}$ at the contact line based on [13]. So $\mathbf{E}_{cl}\mathbf{Oh}$ has the meaning of a competition parameter, a measure of the competition between evaporation and capillarity at the contact line. Analogously $\mathbf{E}\,\mathbf{Oh}$ is the measure of the competition between evaporation and capillarity at the interface, interpreting $J/\rho_0 = U_{ev}$ as the evaporative velocity scale at the interface.

The field equations in the vapor phase in terms of dimensionless numbers read

$$\frac{\partial \hat{\mathbf{u}}_v}{\partial \hat{t}} + \hat{\mathbf{u}}_v \cdot \hat{\nabla}\hat{\mathbf{u}}_v - \mathbf{Oh_v}\hat{\nabla}^2\hat{\mathbf{u}}_v + \hat{\nabla}\hat{P}_{v,hyd} = 0, \tag{3.68}$$

$$\hat{\nabla} \cdot \hat{\mathbf{u}}_v = -\hat{c}, \tag{3.69}$$

$$\frac{\partial \hat{T}_v}{\partial \hat{t}} + \hat{\mathbf{u}}_v \cdot \hat{\nabla}\hat{T}_v - \frac{\mathbf{Oh_v}}{\mathbf{Pr_v}}\hat{\nabla}^2\hat{T}_v = \frac{1}{\mathbf{Ptd_v}}\frac{\partial \hat{P}_v}{\partial \hat{t}}, \tag{3.70}$$

The expression for $\hat{c}$ becomes

$$\hat{c} = \frac{1}{V_v}\int_{\Gamma_I}\hat{\mathbf{u}}_l \cdot \mathbf{n} + \left(\mathbf{R}_\rho - 1\right)\mathbf{E\,Oh}\,\mathrm{d}\Gamma_I, \tag{3.71}$$

The boundary conditions at the interface from the side the vapor are

$$\hat{\mathbf{u}}_v = \hat{\mathbf{u}}_l + \left(\mathbf{R}_\rho - 1\right)\mathbf{E\,Oh}\hat{j}\mathbf{n}, \tag{3.72}$$

$$\hat{T} = \hat{T}_S(\hat{P}_v), \tag{3.73}$$

The mass flux at the interface is

$$\hat{j} = -\frac{1}{\mathbf{Ev}}\partial_\mathbf{n}\hat{T}_l + \frac{1}{\mathbf{Ev_v}}\partial_\mathbf{n}\hat{T}_v = -\partial_\mathbf{n}\hat{T}_l + \frac{1}{\mathbf{Ev_v}}\partial_\mathbf{n}\hat{T}_v, \tag{3.74}$$

taking into account $\mathbf{Ev} = 1$, whereas the expression for the evaporative flux at the contact line in dimensionless numbers

$$\hat{j}_{cl} = \hat{K}^{-1}\hat{T}_w \tag{3.75}$$

remains the same.

The energy equation in the wall becomes terms of dimensionless numbers

$$\frac{\partial \hat{T}_w}{\partial \hat{t}} - \frac{1}{\mathbf{Pe_w}}\hat{\nabla}^2\hat{T}_w = 0, \tag{3.76}$$

3.3.3 Scaling concept and ranges

The following section has the goal to identify the important dimensionless numbers through which the variations of the reorientation characteristics are described in Chapter (5). For the purpose the importance of the dimensionless numbers regarding the motion of the liquid and the interface is discussed. Alongside with this the ranges of the variation of the relevant numbers in Table 3.1 are presented for the variation of the experimental conditions. All values of the relevant dimensionless numbers ($\mathbf{Ev}_v$, $\mathbf{E}$, $\mathbf{E}_{cl}$,$\mathbf{Oh}$, $\mathbf{E\,Oh}$ and $\mathbf{E}_{cl}\,\mathbf{Oh}$) in Table 3.1 alongside with the scales of mass flux (J and J_{cl}), velocity U_{puR} and the thickness of the interface interface boundary layer δ_T were calculated according to the definitions in the previous section through the values of the material properties of the fluid phases at the saturation temperature T_S corresponding to the initial vapor pressure P_v of each experiment. These values of the material properties are given in Table A.27-A.31.

The variation of the axial wall temperature gradients $\Delta T_w/\Delta Z$ is connected with a significant variation of the characteristic temperature differences Θ, Θ_w and Θ_v due to the the heating procedure. This in turn is connected with a significant variation of the corresponding dimensionless numbers - $\mathbf{E}$, $\mathbf{E}_{cl}$, $\mathbf{Ev}_v$ and $\mathbf{Ptd}_v$. All other dimensionless numbers are unrelated to Θ, Θ_w and Θ_v. So their variation is attributed to the temperature dependence of the material properties and therefore insignificant.

$\mathbf{Ptd}_v$ does not appear in an equation related to the liquid flow. It appears in the energy equation of the vapor. In contrast $\mathbf{E}$ and $\mathbf{E}_{cl}$ appear in equations directly related to the liquid and interface motion. $\mathbf{E}$ appears in the mass balance boundary condition at the interface Eq. (3.63) which relates the dimensionless velocities of the interface and the liquid to the dimensionless interfacial mass flux. In an analogous way $\mathbf{E}_{cl}$ appears in the contact line boundary condition, Eq. (3.65) or Eq. (3.66), relates the dimensionless velocities of the contact line and the liquid to the dimensionless mass flux at the contact line.

$\mathbf{Ev}_v$ does not appear directly in an equation related to the motion of the liquid or the interface, however, the dimensionless mass flux $\hat{j}$, which appears in Eq. (3.63), depends on $\mathbf{Ev}_v$ through Eq.(3.74). So the influence of $\mathbf{Ev}_v$ on the liquid flow is indirect. This is the way the interfacial phase change affects the liquid and interface motion.

Therefore the variation of the reorientation characteristics from the experiments can be regarded as dependencies on the variation of $\mathbf{E}$, $\mathbf{E}_{cl}$ and $\mathbf{Ev}_v$. This conclusion is also backed by the variation of the values of the dimensionless numbers in Table 3.2 corresponding to the variation of $\Delta T_w/\Delta Z$ the during the experiments. The dimensionless numbers describing the capillary-driven motion of the liquid and the interface are $\mathbf{Oh}$ and $\mathbf{We}$. The value of $\mathbf{Oh}$ fluctuates insignificantly whereas $\mathbf{We} = 1$. $\mathbf{E}_{cl}$ varies more than one magnitude for all experiments. Principally the variation of $\mathbf{E}$ exceeds one magnitude, too, where the experimental data allows the calculation. $\mathbf{Ev}_v$ vary even more than two magnitudes. In this way Table 3.2 illustrates my main assumption of the reorientation and its mathematical modeling: capillary-driven flow (essentially constant) with interfacial phase change (significantly varying).

Both $\mathbf{E\,Oh}$ and $\mathbf{E}_{cl}\,\mathbf{Oh}$ have the meaning of the measure of competition between evaporation and capillarity as mentioned. Simultaneously they are also measures of evaporation, given the almost constant value of $\mathbf{Oh}$, through which my results can be compared to the literature. Therefore I consider these products of dimensionless number as the actual dimensionless numbers which are representative for my ex-

Table 3.2. Variation of the important dimensionless numbers for the reorientation. The values of LAr with geometry A are denoted with (A) and with geometry B are denoted with (B).

	LAr	LCH$_4$	LNe	LH$_2$	
Ev	1	1	1	1	
Ev$_v$	0.00-2.47 (A) / 0.56-1.60 (B)	0.02-2.89	0.0005-0.097	36.3	$\times 10^6$
E	0.00-1.95 (A) / 1.19-1.60 (B)	0.23-2.95	0.01-1.60	0.88-4.32	$\times 10^{-2}$
E$_{cl}$	0.16-2.36	0.14-2.34	0.06-3.58	0.001-3.20	$\times 10^{-1}$
Oh	3.65-3.98	3.00-3.41	2.97-3.10	2.19-2.28	$\times 10^{-4}$
E Oh	0.00-7.16 (A) / 4.36-5.83 (B)	0.80-9.24	0.03-4.76	2.00-9.46	$\times 10^{-6}$
E$_{cl}$ Oh	0.64-5.16 (A) / 4.71-8.61 (B)	0.47-7.33	0.19-10.80	0.002-7.19	$\times 10^{-5}$
We	1	1	1	1	

perimental results based on [13] where the competition parameter $\hat{C}^*$ appears as a dimensionless number in the corresponding equation, as already mentioned. As the third relevant dimensionless number I consider **Ev$_v$**.

Here I must point out that if the mass flux at the interface is modeled according to Eq. (2.17), assuming an interface temperature equal to the saturation temperature, the motion of the liquid and the interface would be influenced by the interfacial phase change only at the contact line. At the interface the mass flux would be zero, as already remarked in Section. (2.3), so that **E** would not arise in Eq. (3.63)). Accordingly the variation of the reorientation characteristics would be attributed solely to the phase change at the contact line and would depend only on **E$_{cl}$**, since **Ev$_v$** would not arise from the non-dimensionlization of Eq. (2.17) at all. However, the mass flux scale at the interface J varies more than one magnitude for the variation of $\Delta T_w/\Delta Z$ for LCH$_4$, Tables A.6 - A.10 in Section A.1.2. In the same time mass flux scale at the contact line J_{cl} varies more than one magnitude, too, Table A.6 - A.10. So the assumption of a zero mass flux seems to me unjustified and therefore I adhere to the modeling of the mass flux at the interface according to Eq. (3.29).

3.4 Step response of linear systems of second order

The step response can be assumed as the basic underlying mechanism describing the reorientation process of the liquid-vapor interface after the step reduction of gravity in terms of system dynamics. Some surface wave effects and wall boundary layer effects are additional to that and can be regarded as a superimposed on the step response. Therefore I will present here the quantitative characteristics of the step response of a second order linear system and their application to the reorientation.

There is a clear connection between the step response model and the approach by [50] and [29] used for the isothermal reorientation on the one side, and the sloshing models by [4] and [38] on the other side.

The equivalent mechanical model of the step response is a mass-spring-damper model. The mathematical description of such a equivalent model was given by [50] and a graphical representation of the model was given by [29]. A mass-spring model was presented by [4] for the lateral sloshing modes due to harmonic excitation whereas in [38] general mass-spring-damper models were presented for lateral translational and pitching excitation. The natural frequency of the sloshing modes are given by the square of the ratio between spring constant and the mass of the equivalent mechanical model corresponding to the mode.

Accordingly there is a connection between the equations of liquid motion corresponding to these models - in each case it is a linear differential equation of second order. The step response model I present is the non-homogenous generalization of the equation used by [50] and [29] through a step force input. The equations used by [4] and [38] account for the harmonic excitations inducing the corresponding sloshing modes.

The advantage of using this step response model is that an explicit analytical solution exists. Based this solution an explicit relation between the natural (undamped) frequency of the oscillation and the steady-state deflection exists. Applying this to the oscillation of the center point Z_c and using its dependence on the steady-state contact angle by Eq. (3.11) gives an explicit analytical dependence of this frequency on the contact angle for arbitrary values of the contact angle. In contrast the modal analysis in [4] and [38] based on the potential theory gives an explicit solution only for the value $\pi/2$ of the contact angle.

However, both approaches show an increasing frequency with the increasing contact angle. This trend is observed for the interval $36°$ to slightly above $\pi/2$ for the first axial sloshing mode $(0, 1)$ by the investigation by [4]. This mode showed the lowest frequency of all axial modes in this investigation. It was noted by [29] that the frequency of the first axial sloshing mode would be the dominant mode regarding the reorientation oscillations since all the other axial modes would be damped more quickly in the presence of viscosity due their higher frequency. Accordingly this frequency can be interpreted as the lowest possible of the experimentally observed reorientation oscillations according to [50] and [29]. Based on [4] this frequency ω_{01} can be represented by the square of the ratio between spring constant and the mass of the equivalent mechanical model corresponding to the axial $(0, 1)$ mode. The step response I present thereafter provides such an equivalent model.

Since by the reorientation the transition from the initial equilibrium to the final equilibrium undergoes a series of characteristic damped oscillations, it is natural to approximate the step response of the liquid-vapor interface to the step reduction of gravity with a step response approach, and not just with damped oscillations. This is expressed also by the major trait of the linear step response of second order - the inverse proportionality of the square of the natural frequency of the oscillations to the steady-state deflection. Similarly an increase of the oscillation frequency is accompanied with a decrease of the inferred steady-state deflection for both the center point and the contact line the corresponding variation of the wall gradients in the experiments treated in this work. A deviation of the experimental data from the this proportionality should be attributed to the non-linear nature of the reorientation due to the above mentioned surface wave effects and wall boundary layer effects. A comparison between the above mentioned proportionality relation and the experimental data is done in Chapter 5.

3.4.1 General characteristics of the step response

For the treatment of the reorientation I propose an equivalent mechanical model of the liquid motion based on the equivalent models presented by [29] (Abbildung 3.6), [4] (Fig.7), [38] (Figure 5.2) and [58]. The model consists of a linear spring with a spring constant (stiffness) k, a moving mass m, a linear damper with a viscous damping constant c and a rigid mass m_0 as shown in Fig. 3.2, all connected along the axis of a cylinder geometry. The moving mass m represents the portion of the liquid participating in the liquid/interface motion during the reorientation whereas m_0 represents the motionless portion of the liquid. For the time $t < 0$ the system is in a equilibrium and at rest. At $t = 0$ the system is subjected to a step disturbance with a force described by $F(t) = a\,u_s(t)$ where a is a constant input with dimensions of a force and $u_s(t)$ is the dimensionless unit step function. The force F is applied at the moving mass m representing the step reduction of gravity. The displacement of m from its equilibrium position with regard of the cylinder is Z. So this is a linear second order system which I assume to be underdamped as well [58].

The equation of motion (slosh equation) for the equivalent mass m then reads

$$m\ddot{Z} + c\dot{Z} + kZ = F(t) = a\,u_s(t) \tag{3.77}$$

where

$$u_s(t) = \begin{cases} 0 \,, t = 0 \\ 1 \,, t > 0 \end{cases} \tag{3.78}$$

which implies that only the moving equivalent mass m exhibits motion under the action of F. Since I define the excitation force F explicitly, and not just the excitation as by [4] and [38], it is directly included in the right hand hand side of the above equation. The initial conditions for Eq. (3.77) are $Z(0) = 0$ and $\dot{Z}(0) = 0$. As already mentioned [50] and [29] used the above Eq. (3.77) with $a\,u_s(t) = 0$ as a substitution

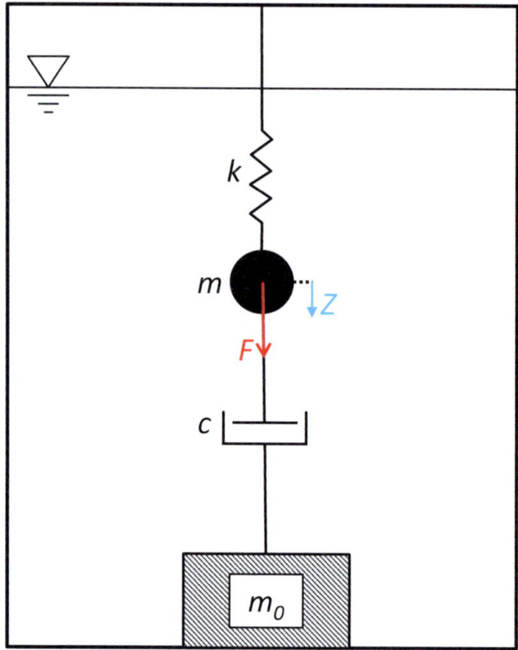

Fig. 3.2. Sketch of the equivalent mechanical model of the liquid motion during the reorientation as a spring-mass-damper system subjected to a step input.

mechanical model of the oscillations during the isothermal reorientation. s If one defines the natural (undamped) frequency

$$\omega_n = \sqrt{\frac{k}{m}} \tag{3.79}$$

and the damping ratio

$$\zeta = \frac{c}{2\sqrt{km}} \tag{3.80}$$

then Eq. (3.77) can be written in the form

$$m\ddot{Z} + 2\zeta\omega_n\dot{Z} + kZ = a\,u_s(t) \tag{3.81}$$

which is mathematically identical with the sloshing equations Eq.(5.5b) and Eq.(5.6) in [38] of the equivalent mechanical models corresponding to the sloshing modes by purely translational lateral excitation with non-zero damping. The term due to the excitation in the above equation is $a\,u_s(t)$. Potential theory and small-amplitude oscillation was employed by [38]. These equivalent mechanical models are mass-spring-damper(dashpot) systems connected to the container walls perpendicular to the container axis. The natural frequency and the damping ratio of these mass-spring-damper systems are given by relations identical with Eq. (3.80) and Eq. (3.80).

Using the same approach as [38] it was stated by [4] that the liquid can be represented by a mechanical model consisting of a mass-springs systems corresponding to the different sloshing (m,n) modes by zero damping. The mass of of such a system is denoted as m_{mn} and termed as sloshing mass. This is the liquid mass participating in the motion by the respective sloshing mode. Using equations corresponding to Eq. (3.81)[4] investigated the dependence of the ratio between m_{mn} and the total liquid mass on the liquid height and static contact angle for m_{11}, m_{12} and m_{13} by lateral harmonic excitation. So in the general case based on [4] the relation would be valid

$$\omega_{nm} = \sqrt{\frac{k_{mn}}{m_{mn}}} \tag{3.82}$$

where ω_{nm} is the natural frequency of the sloshing mode, k_{mn} is the spring constant of the equivalent mechanical model and m_{mn} is the corresponding sloshing mass. For the axisymmetric sloshing mode $(m = 0, n = 1)$ this would give

$$\omega_{01} = \sqrt{\frac{k_{01}}{m_{01}}} \tag{3.83}$$

so that an equivalent mechanical model with ω_{01}, k_{01} and m_{01} by a stepwise axial excitation would exactly correspond to the step response representation of reorientation given by Fig. 3.2 and Eq. (3.77).

The analytical solution of the equation of motion Eq. (3.81) can be written as

$$Z(t) = \frac{a}{m\omega_n^2}\left[1 - \frac{e^{-\zeta\omega_n t}}{\sqrt{1-\zeta^2}}\sin\left(\omega_n\sqrt{1-\zeta^2}\,t - \arctan\frac{1-\zeta^2}{\zeta}\right)\right] \tag{3.84}$$

describing a damped oscillatory motion staring from $x(0) = 0$ and reaching a steady state value of

$$Z_{ss} = Z(\infty) = \frac{a}{m\omega_n^2} \tag{3.85}$$

which is the key feature of the linear step response of second order, as already mentioned. The quantities ω_n and ζ are the undamped natural frequency and the damping ratio of the oscillatory motion, respectively.

The above relation can be used in combination with Eq. (3.11) to give an explicit analytical expression of the dependence of the natural frequency of oscillation of the center point on the steady-state contact angle γ_{ss}

$$\omega_n \propto Z_{css}^{-1/2} \propto \left\{ -\frac{R}{\cos \gamma_{ss}} \left[1 - \frac{2(1 - \sin^3 \gamma_{ss})}{3 \cos^2 \gamma_{ss}} \right] \right\}^{-1/2} \tag{3.86}$$

So if Z_{ss} is varied in Eq. (3.85) with a and m kept constant then the ω_n will vary accordingly

$$\ln \frac{Z_{ss,1}}{Z_{ss,2}} = 2 \ln \frac{\omega_{n,2}}{\omega_{n,1}} \tag{3.87}$$

where $Z_{ss,1}$ and $Z_{ss,2}$ are the steady-state deflections of the original and the varied step response, respectively. The corresponding frequencies are $\omega_{n,1}$ and $\omega_{n,2}$. Equation (3.87) is compared in Chapter 5 to the correlation inferred from the experimental data.

The condition for the system to be underdamped means that $0 \leq \zeta < 1$. The actual frequency of the damped oscillation, the damped natural frequency ω_d, appears in the argument of the sine function of the solution Eq. (3.84) and is

$$\omega_d = \omega_n \sqrt{1 - \zeta} \tag{3.88}$$

to which a corresponding period $T_d = 2\pi/\omega_d$ is assigned. The quantity ω_d is the frequency of the free oscillations of the system about the steady state x_{ss} after being displaced from the equilibrium by the step input. As it can be seen ω_d is dependent only on the system properties and independent from the input.

In this manner the four constants m, c, k and a determine the solution of the equation of motion completely. On the other hand the three parameters ω_n, ζ and x_{ss} determine the solution completely, too.

The relation between frequency of the oscillations and their steady-state amplitude, expressed by Eq. (3.85), is a principal feature of the linear step response of 2nd order, which is also qualitatively compatible with the other types of frequency relations of oscillations based on [4], Eq. (2.38)-(2.39) and Eq. (2.40)-(2.43), in the literature treating the isothermal reorientation which are presented in Section (2.6.1).

3.4.2 Transient characteristics of the step response

I am are interested in present work in the quantitative characterization of reorientation of the interface subjected to a variation of its boundary condition. Since I have assumed the step response to be the basic underlying mechanism, the characteristics of the reorientation process can be identified as the transient characteristics of the step response of a linear system of second order. The definitions of these transient characteristics follow.

I have already presented in the previous Section (3.4.1) the full set of characteristics determining the step response signal - the value of the steady-state response x_{ss}, the undamped natural frequency ω_n and the damping ratio ζ. However, there are also other characteristics of the step response signal describing the transient part of it. Some of these characteristics were used in some way for the investigation of the interface oscillations after the step reduction of gravity as already mentioned, e.g. the first pass through the final equilibrium value t_{s1} in Eq.(2.35) by [72] and the settling time t_s in Eq. (2.36) by [82], however, not as functions of the undamped frequency and the damping ratio of a step response. That is why I present the mathematical relations of theses characteristics in terms of ω_n and ζ for the step response of an underdamped second order system described by Eq. (3.84) as derived in [58] and [59], omitting the actual derivation.

The peak time t_p is needed by the signal to achieve its maximal deflection (or peak value x_p) from its initial value. Therefore one can obtain the peak time by differentiating $x(t)$ in Eq. (3.84) with respect to time t and letting this derivative equal zero. This requires that t_p is

$$t_p = \frac{\pi}{\omega_n \sqrt{1 - \zeta^2}} = \frac{\pi}{\omega_d} \tag{3.89}$$

In other words the peak time corresponds to one half-cycle of the frequency of damped oscillation ω_d.

The peak value Z_p of the transient response occurs at the the peak time t_p. Substituting Eq. (3.89) in Eq. (3.84) gives for Z_p

$$Z_p = \frac{a}{m\omega_n^2} \left[1 - \frac{e^{-\zeta/\sqrt{1-\zeta^2}}}{\sqrt{1 - \zeta^2}} \sin\left(\pi - \arctan \frac{1 - \zeta^2}{\zeta} \right) \right] \tag{3.90}$$

which considering Eq. (3.85) delivers the relation

$$Z_p = Z_{ss} \left[1 - \frac{e^{-\zeta/\sqrt{1-\zeta^2}}}{\sqrt{1 - \zeta^2}} \sin\left(\pi - \arctan \frac{1 - \zeta^2}{\zeta} \right) \right] \tag{3.91}$$

rendering the peak value Z_p as proportional to the steady state value Z_{ss} of the step response multiplied by a function of the damping ratio ζ.

The rise time of a underdamped second order system t_r is usually the time needed for the signal to reach the steady state Z_{ss}.

$$t_r = \frac{1}{\omega_d} \arctan \frac{\sqrt{1 - \zeta^2}}{\zeta} \tag{3.92}$$

The maximum overshot M_p occurs at the peak time $t_p = \pi/\omega_d$. Using Eq. (3.91) one gets

$$M_p = Z_p - Z_{ss} = -Z_{ss} \frac{e^{-\zeta/\sqrt{1-\zeta^2}}}{\sqrt{1 - \zeta^2}} \sin\left(\pi - \arctan \frac{1 - \zeta^2}{\zeta} \right) \tag{3.93}$$

Usually the maximum percentage overshoot is used

$$M_{p,perc} = \frac{Z_p - Z_{ss}}{Z_{ss}} \times 100\% = -\frac{e^{-\zeta/\sqrt{1-\zeta^2}}}{\sqrt{1 - \zeta^2}} \sin\left(\pi - \arctan \frac{1 - \zeta^2}{\zeta} \right) \tag{3.94}$$

The settling time t_s can be defined as

$$t_s = \frac{4}{\zeta \omega_n} = 4T \qquad (3.95)$$

when the signal is within the $\pm 2\%$ tolerance around the steady-state Z_{ss}. Note that the settling time defined in this way is 4 times the time constant T which is inversely proportional to the product of the damping ration and the undamped frequency. The time constant T is chracteristic time (a time scale) for the speed of the decay of the damped oscillations. The curves $Z_{ss}(1 \pm \exp(-\zeta \omega_n t)\sqrt{1-\zeta^2})$ which are the envelope curves of the transient signal exhibit exactly the time constant $1/\zeta \omega_n$.

4 Materials and methods.

The following chapter describes the instrumentation and the methodology used for performing the experiments studied here, the evaluation and the error analysis of the acquired data.

The essential components for conducting the experiments is a right circular cylinder (Fig. 4.3 and Fig. 4.4) of fused silica, a ring heat foil on the outside top part of the cylinder, a high-speed recording CCD camera capturing the liquid-vapor interface through an endoscope, an illumination unit, a series of temperature sensors measuring the temperatures of the experiment liquid and its vapor as well as of the cylinder wall at specified positions, and finally a pressure transducer measuring the vapor pressure. The experiment cylinder, partly filled with the experimental liquid, is housed within a cryostat setup (Fig.4.1 and Fig. 4.2) ensuring the necessary cryogenic temperatures. Except for the pressure transducer and the CCD camera (and the illumination unit in the second setups) all other essential components of the experimental setup are accommodated within the cryostat setups. Prior to the step reduction of gravity the cylinder wall above the interface is heated through the ring heat foil, achieving different target values of the wall temperature gradients towards the interface. After the end of the heating a step reduction of gravity follows, the cold cryogenic liquid starts to rise along the warmer cylinder wall and a reorientation process of the free surface through a series of damped oscillations is established.

The experiments were performed at the Bremen Drop Tower facility. The actual 110 m drop distance provides reduced gravity conditions for about 4.7 seconds. The sudden transition from 1g to μg conditions is initiated by the release of a drop capsule in which the experimental setup is accommodated inside the evacuated drop tube. The drop capsule is equipped with a computer acquiring the experimental data during the drop. The tube vacuum of approx. 10 Pa enables residual accelerations of less than 10^{-5}g during the drop.

4.1 Cryostat setups. Operation and experiment preparation.

Two specially designed cryostat setups were implemented for the experiments. The cryostat setups allow different cryogenic liquids to be used for cooling and experiment purposes. The setups are depicted in Fig. 4.1 and Fig. 4.2. The first cryostat setup housed experimental cylinder A shown in Fig. 4.3, since it is referred to as cryostat setup A. The second cryostat setup and experimental cylinder are referred to as cryostat setup B and cylinder B, respectively. Both experimental cylinders are described in the next Section 4.2 whereas a discussion of the cryostat setups follows and their operation. The actual procedure for the execution of the experiments is described in Section 4.8.

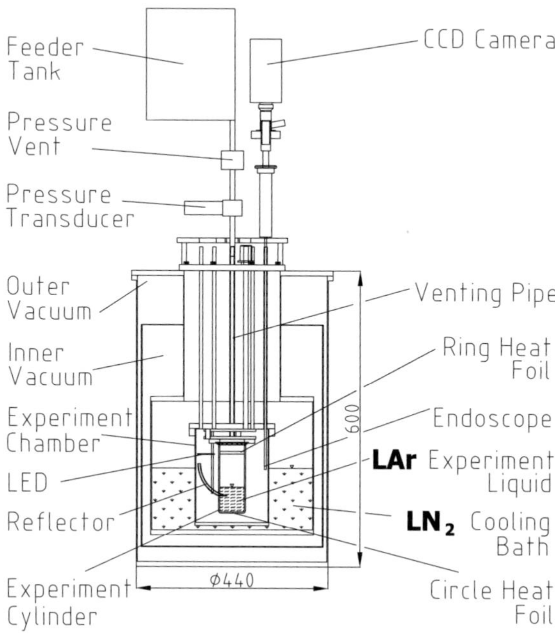

Fig. 4.1. Cryostat setup A with the experimental cylinder A.

4.1.1 Cryostat setups

Both setups share general features. The cryostat has outer dimensions of ⌀ 440 mm and 600 mm with a double wall structure of stainless steel providing outer vacuum insulation for the experiment cylinder. The insulation eliminates the convective and the conductive heat transfer into the cryostat and only radiative transfer between the walls can occur. Radiation shielding is implemented under the top of the inner wall structure to minimize the radiative heat transfer towards the experimental chamber. Additionally foam insulation is utilized to minimize the parasitic conductive heat

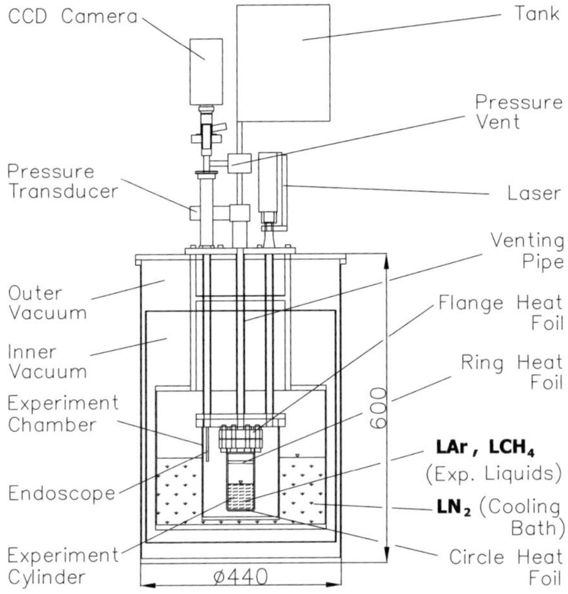

Fig. 4.2. Cryostat setup B with the experimental cylinder B.

flow from the environment into the experiment chamber. Venting pipes extend from the ambient region (around 300 K and 1.013 kPa) to the cryogenic region. The experimental cylinder itself is mounted through a connection, o-ring (setup A) or a flange (setup B), on the ends of the pipes in the cryogenic region. In this way the space occupied by the vapor of the experiment liquid extends from the cryogenic region to the ambient region, which will be discussed in detail in Sec. (4.2).

The experiment chamber is the hollow space with a coaxial cylindrical geometry with an inner diameter of 150 mm and an outer diameter 180 mm providing inner vacuum insulation for the experiment cylinder. It contains besides the experiment cylinder also the illumination unit (setup A), the reflector (setup A) or the reflecting surfaces (setup B) and the endoscope. It was evacuated during the experiment operation, down to 10 Pa for the experiments with setup A and down to 0.4 Pa (LAr and LCH$_4$) or 0.02 Pa (LNe and LH$_2$) for these with setup B, in order to eliminate the convective heat fluxes towards the experiment cylinder and the condensation on the endoscope or the cylinder wall. The experiment chamber is partially submerged into the cooling bath, ensuring so the necessary saturation temperatures for the condensation and storage of the experimental liquid.

Cooling bath of LN$_2$ at $T \approx T_{sat} = 77.35$ K and $P = 1013$ hPa, was used in the setup A (approx. 20 L) and for the experiments with liquid argon and methane with setup B (a typical depth of 50-90 mm). For the experiments with liquid neon and liquid hydrogen conducted with setup B LHe (at $T \approx T_{sat} = 4.2$ K and $P = 1013$ hPa) was used as coolant.

A key feature of both setups is the thermal decoupling of the experiment liquid and its vapor from the cooling bath. This is required in order to eliminate the freezing risk for both the liquid and the vapor due to the lower saturation temperature of the cooling bath compared to the experiment liquid. The corresponding precautions are realized in different ways in the two setups.

In setup A a specially designed separator flange, mechanically connecting the experiment chamber and the cylinder, enables the thermal decoupling of the cooling bath liquid from the experiment liquid and prevents from the freezing of the latter one. The formation of vapor ice plugs in the venting pipe and the destruction of the vapor stratification in the experiment cylinder are avoided by achieving a higher temperature of the top plate of the experiment chamber than the saturation temperature of the experiment liquid through the partial coverage of the experiment chamber by the cooling bath with LN. The lowered cooling capacity of the bath due to its lowered fill level can not compensate the parasitic heat flow from the environment to which the top plate is exposed to. Given that and the big thermal inertia of of the plate small wall temperature gradients in the experiment cylinder towards the liquid-vapor interface arise and are sustained.

In setup B the thermal decoupling of the vapor by utilizing a double wall structure for the venting pipes. This means that the actual venting pipes which are connected to the experimental cylinder are placed within pipes with a bigger diameter. So a hollow coaxial space between the two types of pipes arises which is connected to the space of the experiment chamber. In this way this hollow space in between the pipes can be either evacuated or pressurized together with the chamber. When evacuated there is practically no thermal contact by conduction or convection of the experimental liquid and the vapor with the bath which eliminates any risk of freezing or de-stratification. Apart from this a heat foil is attached to the flange flange, the flange heat foil in Fig.4.2, of the experiment cylinder which allows the flange together with neighboring parts of the experiment cylinder and the venting pipes to be heated if needed. This alongside with big thermal mass of the flange allows to avoid the aforementioned risk.

4.1.2 Operation and experiment preparation.

Prior to the experimental campaigns an accurate cleaning of the inner surface of the experimental cylinders (Fig. 4.3 and 4.4) is ensured through multiple ultrasonic baths in isopropanol.

The experiment preparation at the beginning of the experimental campaign starts with the evacuation to approx. 5×10^{-3} Pa of the experiment cylinder by an opened pressure vent. It is followed by the the cool down of the cryostat through the direct filling with LN_2 or LHe. Continuous monitoring of the temperature and the pressure in the experiment cylinder is carried out at a sampling rate of 1 Hz. The temperature at the inner side of the experiment chamber top plate is monitored through an additional sensor. After reaching the saturation temperature of the cryogenic liquid in the experiment cylinder the pressure vent is opened and the flow rate of the condensing gas is regulated through an expansion valve.

The required fill level h_0 of the experiment liquid is achieved by considering the interaction of the pressure, temperature and condensation rate of the gas in the experiment cylinder with the cooling bath fill level. It is visually controlled through the endoscope. The pressure vent and the expansion valve are closed immediately by reaching the required fill level so that this constant fill level is ensured for all experiments under the conditions of the closed system of the experiment cylinder, the venting pipe and the respective portions of the tubings above it up to the pressure vent. Management of the residual pressure in the experiment chamber and the sustainment of the LN_2/LHe cooling bath fill level enable maximum stable thermodynamical conditions in the setup. During the actual experiment procedure it is possible to achieve almost the same starting conditions prior to the application of the axial wall temperature gradients by heating through the bottom circle heat foil or by cooling through the management of the experiment chamber residual pressure.

4.2 Experimental cylinders and liquids

As already mentioned two experimental cylinders were implemented: cylinder A, depicted in Fig. (4.3), and cylinder B, depicted in Fig. (4.4). Four experimental liquids were investigated - liquid argon with both cylinders as well as liquid methane, neon and hydrogen with cylinder B. This section presents the experimental cylinders with their dimensions and material composition, and material properties of the three interacting phases - liquid, vapor and solid. A very important feature is that the experiment cylinder forms a thermodynamically closed system with the venting pipes and the tubings above them during the experimentation. Details on this follow.

There are common essential features of the cylinders. In both case a right circular cylinder is implemented which is partially filled with the experimental liquid. The cylinder in its main part, interacting with the liquid, is made of fused silica/borosilicate. The fill level of the experimental liquid is at least equal to the inner radius of the cylinder multiplied by a factor of 1.6 in order to avoid the influence of the cylinder bottom on the interface motion, [30]. At a specific height above the interface a heating foil is mounted through which the wall gradients towards the interface are generated. The pressure measurement of the vapor is conducted through a transducer outside of the cryostat. The temperature of the wall and inside the cylinder is measured by sensors positioned there.

The measurement of the liquid, vapor and solid temperatures is conducted point wise in geometry A: a single point in each phase - bulk temperature of liquid and vapor, wall temperature next to them and next to interface, through which the wall temperature gradients of the experiments.

The measurement of the vapor and solid temperatures in geometry B was conducted array wise - temporal development of the vapor temperature distribution (stratification), temporal development of the wall temperature distribution within the expected range of motion of the contact line (and so of the axial wall temperature gradients, too). No measurement of the liquid temperature was performed due to technical difficulties.

The experiment cylinder (Fig. 4.3) is made of fused silica and has the dimensions: inner diameter $R = 25.5$ mm, outer diameter $R_{out} = 28$ mm, and height $H_{out} = 155$ mm. After the reduction of gravity a transient oscillation of the free surface is established by the reorientation of the liquid from the static 1g equilibrium configuration to the μg equilibrium configuration, which cannot be reached within the 4.7 s of microgravity with LAr. The fill level of $h_0 = 59.15$ mm ($\geq 2R$) of the cryogenic liquid is chosen in order to avoid the effect of the bottom on the reorientation process (see [30]) and to enable the observation of the motion of both the center axial point $Z_c(t)$ and the contact line $Z_w(t)$ during the entire μg phase. The space above the free surface is occupied by argon vapor GAr.

At a height $Z_H = 126.4$ mm from the inner bottom of the cylinder a ring heat foil of breadth $b_H = 6$ mm is mounted, through which the axial wall temperature gradients towards the free surface are achieved. At the level of $Z_v = 70$ mm above the free surface and within $r_T = 16.5$ mm from the cylinder axis the vapor temperature T_v is measured. In the same manner at $Z_l = 44.5$ mm beneath the free surface the liquid temperature T_l is measured. The wall temperature is measured at three positions: at the free surface level, T_{wf}, and at the temperature measurement levels in the fluid phases, T_{wv} and T_{wl}, respectively. The vapor pressure is measured by the pressure transducer outside of the cryostat through the venting pipe.

Table 4.1. Material properties at 87.3 K

Material	ρ [kg/m^3]	c_p [J/K-kg]	λ [W/m-K]	μ [Pa-s]	σ [N/m]
fused silica	2200.0	230.0	6.35×10^{-1}	-	-
LAr	1395.4	1117.2	1.30×10^{-1}	2.6×10^{-4}	1.25×10^{-2}
GAr	5.8	565.8	5.70×10^{-3}	7.0×10^{-6}	-

Table 4.2. Experiment cylinder A - dimensions and important distances

H_{OUT} [mm]	R_{OUT} [mm]	R [mm]	h_0 [mm]	Z_l [mm]	Z_v [mm]	r_T [mm]	Z_H [mm]	b_H [mm]
155.0	28.0	25.5	59.15	44.5	70.0	16.5	126.4	6.0

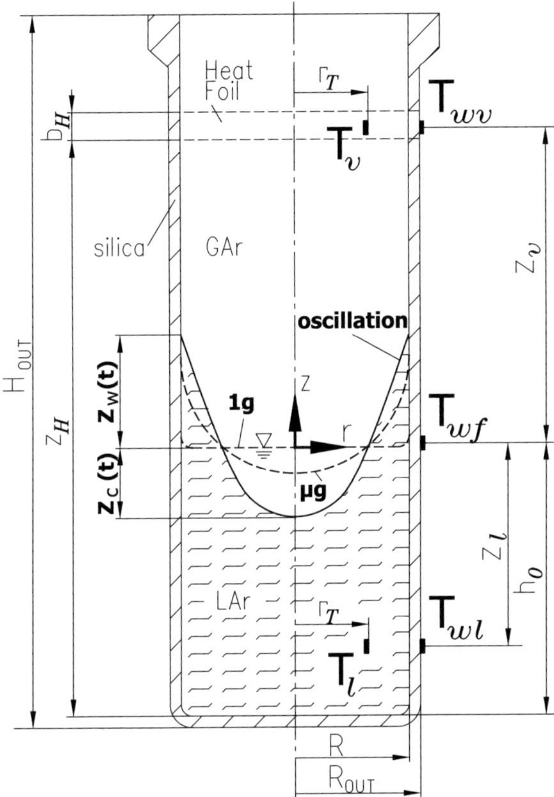

Fig. 4.3. Experiment cylinder A with the positions of the temperature sensors.

After the experiment campaign mass spectrometry measurements of the chemical composition of the experiment vapor yielded the following component values: GN at 44±7 ppm, GO at 8±2 ppm and GHe under the detection limit of 10 ppm. With these values I consider the system to be a single-component system.

The experimental cylinder used for the experiments was a right circular cylinder made of borosilicate glass, similar to the one described in [46]. The cylinder with its dimensions, material composition and temperature sensor instrumentation is shown in Fig. 4.4. Liquid argon and liquid methane were used as experimental liquids and their properties are listed in Table A.26. The range for the Ohnesorge number for the experiments with LAr was between 3.64×10^{-4} and 3.73×10^{-4}, calculated through the NIST database ([49]) with material properties at the saturation temperatures given in Table A.2. The range of the Ohnesorge number for the

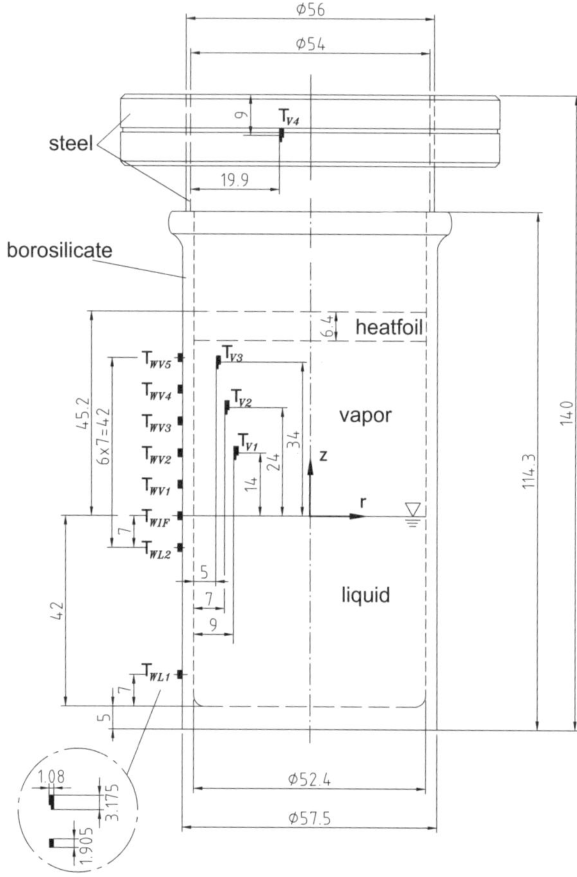

Fig. 4.4. Experiment cylinder B with the positions and the dimensions of the temperature sensors.

experiments with LCH was between 2.96×10^{-4} and 3.41×10^{-4} calculated in the same way from the corresponding values in Table A.3. The filling height was chosen to be greater than the inner radius multiplied by a factor of 1.6 in order to avoid the bottom effects ([30]). The space above the interface was occupied by the cryogenic vapor and beneath it by the cryogenic liquid. The vapor occupies not just the upper portion of the cylinder but the space of the venting pipe and tubes up to the pressure transducer. Thus the total volume of the vapor is 6.45×10^{-4} m^3 $\pm 5 \times 10^{-6}$ m^3. The volume of the vapor in the venting pipe is 0.89×10^{-4} m^3 and its temperature is assumed to increase linearly along 0.4 m from the T_{V4} values in Tables A.2 and A.3 up to the ambiance temperature of around 300 K. The volume of the vapor in the tubes on the top of the cryostat below the pressure vent (see Fig.4.2) is 3.52×10^{-4} m^3 at the ambiance temperature of around 300 K. The cylinder consists of a 114.3 mm long part of borosilicate glass from the bottom up, followed by a 25.7 mm long extension of stainless steel 304 connected to a modified $3^{3/8}$ " CF flange, also made of stainless steel 304. The material properties of the solids are shown in Table A.26.

4.3 Temperature measurement

The temperature measurement was conducted with DT-670 silicon diodes temperature sensors from Lake Shore. The sampling rate during the experiments was 1 kHz. During the preheating the sampling rate was 10 Hz. The sensors accuracy is $\pm$ 0.25 K up to 100 K and $\pm$ 0.5 K above it. The experiments were conducted without measuring the liquid temperature due to technical difficulties. The vapor temperature was measured by the sensors T_{V1} to T_{V4} depicted in Fig. 4.4. The temperature of the outside cylinder wall was measured by the sensors T_{WL1}, T_{WL2}, T_{WIF} and T_{WV1} to T_{WV5}. The positions of all temperature sensors with the sensor dimensions are shown in Fig. 4.4.

4.4 Pressure measurement

The pressure was measured by the ATM analogue pressure transducers, produced by Tetratec Instruments GmbH, situated outside of the cryostat. The measurement range was set to 0 - 2.5 bar. The accuracy within this range is $\pm$ 0.1 % of the end range value, thus 2.5 hPa. The sampling rate of the pressure measurement was 1 KHz.

4.5 Heating equipment

A heat foil of polyimide, model HK 5201 produced by MINCO, was used for the heating process for the achievement of the initial boundary conditions.

4.6 Illumination unit and reflector

The illumination for geometry A an LED panel was implemented in the experiment chamber. For geometry a laser laser with wavelength of 655 nm and 1 W power outside of cryogenic space was used. As a reflecting surface a background reflector behind the cylinder was used whereas in geometry B the experimental chamber surface was covered with aluminum foil (bottom and side walls).

4.7 Visual detection and data evaluation

A CCD camera connected to an endoscope was used for the visual detection of the interface motion. The illumination was realized through the scattering of a laser beam on the bottom of the experimental chamber (Fig. 4.2). The laser (655 nm, 1 W) was mounted outside of the cryostat, a 10 mm diameter endoscope was used.

The visual data evaluation with cylinder B required manual processing of the images due to the difficult illumination conditions. The accuracy of the visual detection is given in Table 4.3.

Table 4.3. Measurement accuracy of the visual detection.

	Z_c	Z_w
	[mm]	[mm]
ΔZ	± 1.08	± 0.66

Two optical effects must be taken into account by the detection of the position of the interface, and in particular of the center point, from the raw images acquired during the experiments since they distort these images in specific ways. The effects are the refraction in the liquid and the fisheye effect of the lens of the endoscope. Refraction distorts in the opposite direction of the fisheye distortion, as it is presented in the following two subsections.

4.7.1 Refraction

The light rays in the experiment propagate in different media - liquid, solid and vacuum. Accordingly a distortion of the image occurs due to the difference in the refraction indices of these media. Due to the effect the apparent dimensions of objects immersed in liquids are bigger than observed in air or vacuum.

The physical law which is required to model the influence of refraction of light is Snell's law. The approach corresponds to the one used by [50] for the investigation of isothermal reorientation through a 2D model of the effects caused by light refraction.

Some assumptions and simplifications are made:

- The lens of the camera is assumed to be equivalent to a geometrical point.
- All light rays are deflected and not reflected.
- It is assumed that the tangential point of the light ray to the interface contour coincides with the center point. This is not the case in reality, the tangential point is slightly above the actual position of the center point. However, this simplification has no influence on the oscillation frequency of Z_w which is investigated in the present work.

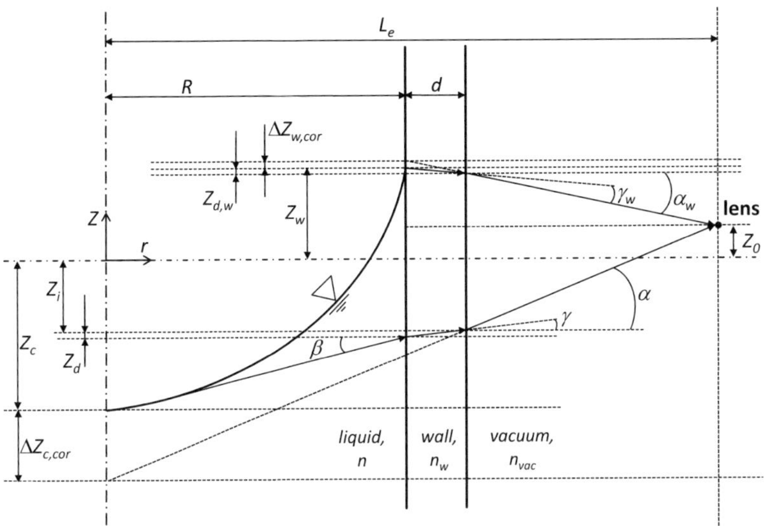

Fig. 4.5. Sketch of the cylinder to model the effect of refraction.

Accordingly the following geometrical representation in Fig. (4.5) is used to model the refraction.

The center of the coordinate system coincides with the position of the center point prior to the step reduction of gravity. The camera lens is positioned at a distance of L_e from the cylinder axis and at as distance of Z_0 above the interface level prior to the step reduction of gravity. The values for L_e were 69.5 mm for geometry A and 57.5 mm for geometry B. The values of Z_0 were slightly different for the individual experiments to ensure best visibility by the varying conditions during the experimental campaigns and these values were determined from the images. R is the inner cylinder radius of geometry A and geometry B with values of 25.5 mm and 26.7 mm, respectively, Fig. (4.3) and Fig. (4.4). The corresponding wall thickness d is 2.5 mm for geometry A and 2.55 mm for geometry B.

Snell's law applied for the center point relates the corresponding angles of the ray propagation from it to the lens - β in the liquid, γ within the cylinder wall and α in the vacuum

$$n \sin \beta = n_w \sin \gamma = n_{vac} \sin \alpha = \sin \alpha \qquad (4.1)$$

where n, n_w and n_{vac} are the refractive indices of the liquid, the cylinder wall and in vacuum, respectively. The refractive index in vacuum is assumed $n_{vac} = 1$. The values of n were taken as 1.23 for LAr, 1.27 for LCH$_4$, 1.09 for LNe, and 1.11 for LH$_2$. The value for n_w was to be 1.51.

Due to refraction the center point after the step reduction of gravity is detected from experimental pictures at an uncorrected (*apparent* or *virtual*) position $Z_{c,u}$

given by $Z_{c,u} = Z_c + \Delta Z_{c,cor}$ where Z_c is the corrected (*real*) position of the center point and $\Delta Z_{c,cor}$ is the correction.

The correction $\Delta Z_{c,cor}$ can be calculated through the following geometrical relations. First,

$$\tan \alpha = \frac{Z_0 + Z_{c,u}}{L_e} \tag{4.2}$$

and besides this

$$\tan \beta = \frac{Z_i}{R} \tag{4.3}$$

where Z_i is the distance in Z-direction between the incidence point of the light ray at the cylinder wall and the real position of the center point Z_c.

The vertical Z_d distance, which the ray travels within the solid, is

$$\tan \gamma = \frac{Z_d}{d} \tag{4.4}$$

where d is the wall thickness of the cylinder.

Apart from this

$$\tan \alpha = \frac{\Delta Z_{c,cor} + Z_i + Z_d}{R + d} \tag{4.5}$$

which can be rearranged as

$$\Delta Z_{c,cor} = (R + d)\tan \alpha - Z_i - Z_d \tag{4.6}$$

Using Snell's law, Eq. (4.2), Eq. (4.3) and Eq. (4.4) the above expression becomes

$$\Delta Z_{c,cor} = (R + d)\frac{Z_0 + Z_{c,u}}{L_e} - R \tan\left(\arcsin\left(\frac{1}{n}\sin\left(\arctan\frac{Z_0 + Z_{c,u}}{L_e} \right) \right) \right)$$
$$- d \tan\left(\arcsin\left(\frac{1}{n_w}\sin\left(\arctan\frac{Z_0 + Z_{c,u}}{L_e} \right) \right) \right) \tag{4.7}$$

where all the information is either extracted from the experimental pictures or is given as geometrical and material properties of the setup.

One can handle refraction in an analogous way regarding the position of the contact line Z_w. The difference is that only refraction within the wall is considered since the ray from the contact line (assumed as a point) towards the lens does not propagate in the liquid, only in the solid in vacuum.

Accordingly the contact line after the step reduction of gravity is detected from experimental pictures at an uncorrected (*apparent*) position $Z_{w,u}$ given by $Z_{w,u} = Z_w + \Delta Z_{w,cor}$ where Z_w is the corrected (*real*) position of the contact line and $\Delta Z_{c,cor}$ is the correction.

Snell's law applied for the contact line is

$$n_w \sin \gamma_w = n_{vac} \sin \alpha_w = \sin \alpha_w \qquad (4.8)$$

Accordingly

$$\tan \alpha_w = \frac{Z_{c,u} - Z_0}{L_e - (R + d)} \qquad (4.9)$$

The vertical $Z_{d,w}$ distance, which the ray travels within the solid, is

$$\tan \gamma_w = \frac{Z_d}{d} \qquad (4.10)$$

On the other hand

$$\tan \alpha_w = \frac{\Delta Z_{w,cor} + Z_{d,w}}{d} \qquad (4.11)$$

giving

$$\Delta Z_{w,cor} = Z_{d,w} \tan \alpha_w - d \qquad (4.12)$$

Applying Snell's law, and using Eq. (4.9) and Eq. (4.10) gives for the correction of the contact line position

$$\Delta Z_{w,cor} = \frac{d(Z_{c,u} - Z_0)}{L_e - (R + d)} -$$
$$d \tan \left(\arcsin \left(\frac{1}{n_w} \sin \left(\arctan \frac{Z_{c,u} - Z_0}{L_e - (R + d)} \right) \right) \right) \qquad (4.13)$$

4.7.2 Fisheye effect

The fisheye effect results from the ultra wide-angle (fisheye) lens of the endoscope. The endoscope lens has an angle of view of 95°. The fisheye effect causes that straight lines which do not go through the center of the lens are pictured as a curved line. Circles around the center of the lens with a given radius, are pictured as circles with a smaller radius, leading to a smaller apparent size of objects than in reality. The optical distortion at the edge of the picture is bigger then in the near of the midpoint of the picture.

In order to determine the distortion due to the fisheye effect a slab of glass with a grid scale on it was used at distance L_e from the endoscope lens. The grid distance was 5 mm. An image of the grid under cryogenic temperatures was taken, however without any liquid. The grid distance of 5 mm corresponded to the 27.8 px from the image. The raw (distorted) distances in vertical direction between the midpoint of the image and the grid-points were measured in px-units, given in Table 4.4, using an edge-detection algorithm in MATLAB. However, the real distances are known, given in Table 4.4 also in px-units, since the grid distance and the resolution are known.

Based on the data in Table 4.4 the following curve fit was applied for the transformation of the vertical pixel coordinates of the raw images into real pixel coordinates along the cylinder axes accounting for the fisheye effect:

$$Z = 3.77 \cdot 10^{-4} Z_{raw}^2 + 1.00588 Z_{raw} + 0.04357 \qquad (4.14)$$

grid point number	Z_{raw} [px]	Z [px]
0	0	0
1	1	1
2	28	28.7
3	55	56.4
4	81	84.1
5	107	111.8
6	132	139.5

Table 4.4. Distortion of the raster image in vertical direction.

4.7.3 Acquisition of the interface position

After removing the distortion of the raw images due to both physical effects, as described in the previous Sections 4.7.1-4.7.2, the acquisition of the interface position can be performed. The acquisition requires the mapping of the pixel coordinates of both the center point and the contact line into real coordinates.

For the center point the mapping of the coordinates is given by

$$Z_c\,[mm] = \frac{Z_c\,[px]}{RES_{cp}\,[px/mm]} \tag{4.15}$$

where RES_{cp} is the image resolution at the center point. The values for RES_{cp} were 5.56 px/mm with geometry A and 5.46 px/mm with geometry B, respectively.

For the contact line the corresponding relation reads

$$Z_w\,[mm] = \frac{Z_w\,[px]}{RES_{cl}\,[px/mm]} \tag{4.16}$$

where RES_{cl} is the image resolution at the contact line. The values for RES_{cl} were 9.33 px/mm with geometry A and 9.44 px/mm with geometry B, respectively.

4.8 Experimental procedure

The experimental procedure and the visual data evaluation are described in detail in [46], so only the main features will be mentioned here. There are two stages of the experiment procedure - the generation of the required value of wall temperature gradients under terrestrial gravity and the following step reduction of gravity. The required gradient value is achieved through a heating of the cylinder wall with the ring heat foil shown in Fig. 4.4. Constant output values of up to 10 W are used for the heating process, which lasts approximately up to 20 min. The heat foil is turned off upon achieving the required temperature gradient value. The wall and vapor temperature distributions after the heating period are depicted in Fig. 4.6 for experiment M17 (see Table A.3). A step reduction of gravity follows due to the drop of the experiment capsule in the Bremen Drop Tower. The vacuum conditions in the drop tower provide for residual accelerations of less than 10^{-5}g during the drop.

The essential feature of the experimental conditions of the experiments was the application of a linear wall temperature distribution

$$T_w(Z) = T_{WIF} + \frac{\Delta T_w}{\Delta Z}Z \qquad (4.17)$$

where $\Delta T_w/\Delta Z$ is the axial wall temperature gradient which was varied during the experiments. Through this a given wall temperature $T_w(Z)$ was established at a height Z above the presumed interface level with a wall temperature T_{WIF}.

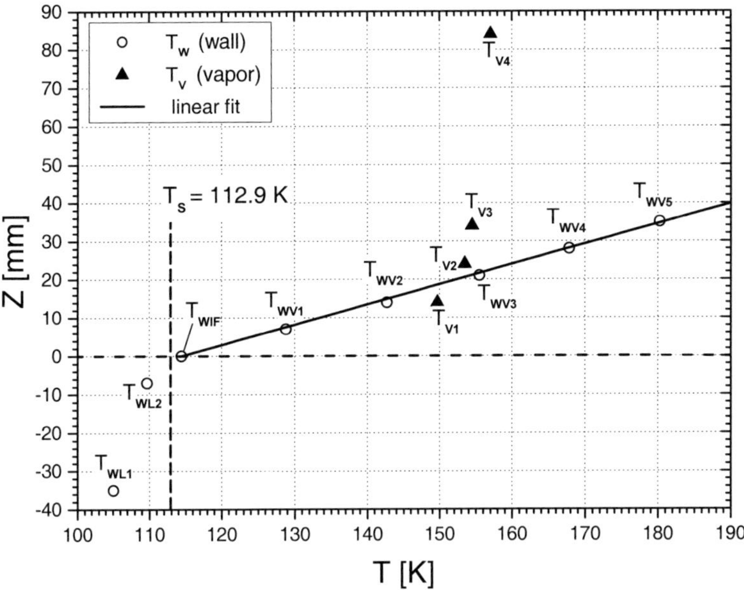

Fig. 4.6. Initial temperature distribution for the experiment M17 with 1.9 K/mm and with LCH$_4$. The solid line represents a linear fit through the wall temperature data with $T_w\,[\mathrm{K}]= 114.4\,[\mathrm{K}] + 1.9\,[\mathrm{K/mm}]Z$ (Z is given in mm).

5 Experimental results and discussion

The structure of the presentation of the experimental results is as follows. First a qualitative description of the behavior of the experimental liquids during the experiments is given in Section 5.1.1. Alongside with this the complete data set from experiment M17 with 1.9 K/mm and LCH_4 is shown as representative case for the majority of the experiments. The graphical representation of the results serves also for the definition of some key variables for the quantitative characterization of the interface reorientation. Then in accordance with the non-dimensional form of the field equations and boundary conditions in Section (3.3.2) and the scaling concept in Section (3.3.3) the dependence of the reorientation characteristics on the values of $\mathbf{E}_{cl}\mathbf{Oh}$, $\mathbf{EOh}$ and $\mathbf{Ev}_v$ is presented in Section (5.2).

5.1 Experiment features

The goal of the experiments was the investigation of the influence of the variation of the value of the non-isothermal boundary condition $\Delta T_w/\Delta Z$ on the interface reorientation. As already mentioned specifically on the motion of two characteristic points was of interest - the center point $Z_c(t)$ and the contact line point $Z_w(t)$. Both motions were observed simultaneously only with liquid argon and methane. Therefore a qualitative description of the general behavior based on the experiments with these two liquids follows.

The comparison of the photos of the final interface configurations with liquid argon and geometry A is presented in Fig. (5.1). The space beneath the contour of the interface is occupied by the cryogenic liquid and space above by vapor. A distinct contact line at this moment can only be observed for $\Delta T_w/\Delta Z = 1.34$ K/mm, whereas the position of the contact line in the other two cases is marked by the kink of the vertical line on the outside back wall of the cylinder due to the contact line itself on the front wall. The experiment with 0.15 K/mm was the only experiment under quasi-isothermal conditions where the contact line could be detected. The contact line motion was thereby quasi-stationary.

A comparison of the photos of final interface configurations LAr and of LCH_4 with geometry B for three different values of $\Delta T_w/\Delta Z$ is presented in Fig. (5.2). Both motions of the contact line and the center point could be thereby observed. The contact line motion was oscillatory.

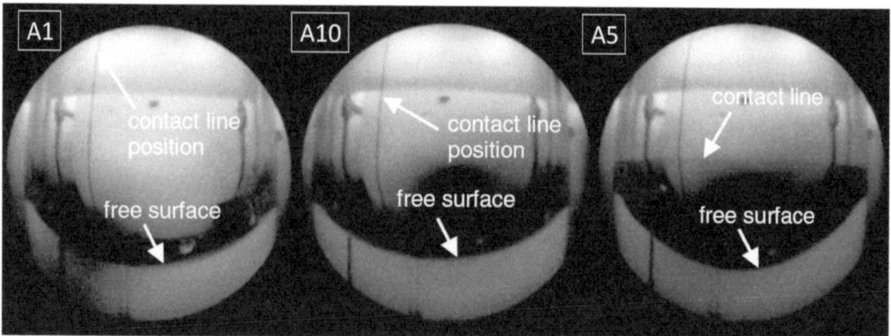

Fig. 5.1. Final interface configuration of LAr during the drop tower experiments A1, A10 and A5 with geometry A, with an increasing wall temperature gradient from left to right.

In general the results from the experiments with LCH_4 and LAr with geometry B confirm the results with LAr and geometry A. The behavior of both liquids was observed to be qualitatively similar concerning the applied boundary condition. Thus the maximum deflection of the center point Z_{cp1} does not vary with the variation of $\Delta T_w / \Delta Z$. With the increase of the value of the wall temperature gradients an increase in the coupling between the motion of Z_w and Z_c was observed. This means that the oscillation frequency of the center point Z_c increases with the increase of $\Delta T_w / \Delta Z$ so that additional extrema are observed. Nucleate boiling at the wall was observed for all experiments except for experiment M24. The pressure was observed to follow the motion of Z_w too. The peaks of the pressure rate coincided with the receding of the contact line. This result suggests that an enhanced evaporation related to the contact line motion takes place on its receding part. In the case of experiment M22 with 2.9 K/mm with LCH_4 the liquid was almost settled at the end of the experiment.

Thus only one experiment was conducted under conditions close to the isothermal situation with $\Delta T_w / \Delta Z = 0$ K/mm. This was experiment M24 with LCH_4 and $\Delta T_w / \Delta Z = 0.2$ K/mm. All other experiments shared certain features, which can be attributed to a greater wall superheat due to the increased $\Delta T_w / \Delta Z$.

5.1.1 Experiment M17 with LCH_4 and 1.9 K/mm

Here I present the full data set from experiment M17 with 1.9 K/mm and LCH_4. The visual conditions enabled the evaluation of both the center point Z_w and the contact line Z_w throughout the experiment. The data represents qualitatively most of the experiments.

The center point Z_c exhibited a damped oscillation with 8 extrema during the experiment as shown in Fig. 5.3. An oscillating motion, also damped, of the contact line Z_w accompanied the motion of the center point (Fig. 5.3). The pressure increased during the experiments in an oscillating manner as seen in Fig. 5.4. The

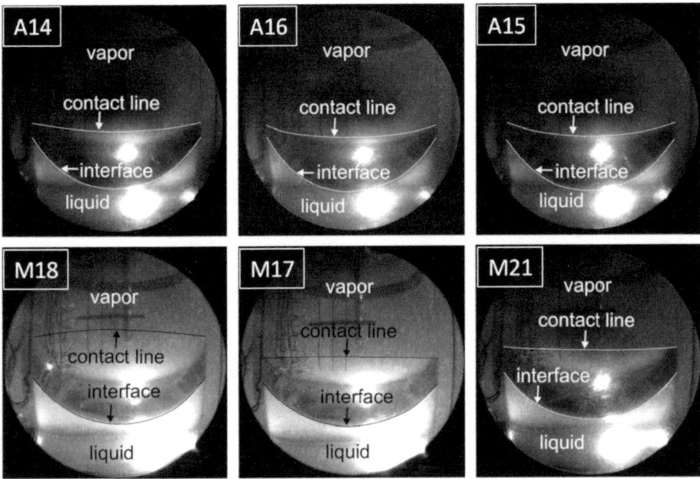

Fig. 5.2. Final interface configurations of LAr during the experiments A14, A16, A15 and of LCH$_4$ during the experiments M18, M17, M21, with an increasing wall temperature gradient from left to right. Two curves were added to the pictures for their better understanding - one for the interface contour and one for the contact line on the front side of the cylinder.

pressure rate dP_v/dt revealed characteristic peaks coinciding with the receding motion, marked gray in Fig. 5.4, of the contact line. These results agree with the results with LAr in [46]. The acquisition of the temperature evolution was possible due to the additional temperature sensors in wall and vapor compared to [46]. The evolution of the wall temperature alongside with the saturation temperature T_S is shown in Fig. 5.5. A temperature decrease is only apparent within the contact line motion, as seen from the curves T_{WV1} and T_{WV2}. The evolution of the vapor temperature is shown in Fig. 5.6. The initial temperature distribution, shown in Fig. 4.6, was significantly changed. The initial temperature difference $T_{WV3} - T_{WV1}$ increased during the experiment and the initial non-linear temperature distribution, given by T_{WV1}, T_{WV2} and T_{WV3}, is replaced by a linear one.

I introduce here through the discussed figures some of the quantities used for the general characterization of the experiments in Sec. 5.2. The period of the half oscillation $T_i/2$ between two consecutive extrema $t_{cp,i}$ and $t_{cp,i+1}$ of the motion of Z_C is defined as $T_i/2 = t_{cp,i+1} - t_{cp,i}$ in Fig. 5.3. The first extremal point of the contact line motion and simultaneously its maximal deflection is defined as Z_{wp1} and its time of occurrence as t_{wp1} in Fig. 5.4. The first peak of the pressure rate and simultaneously its maximum value is defined as dP_v/dt_{pp1}, and its time of occurrence as t_{pp1} in Fig. 5.4.

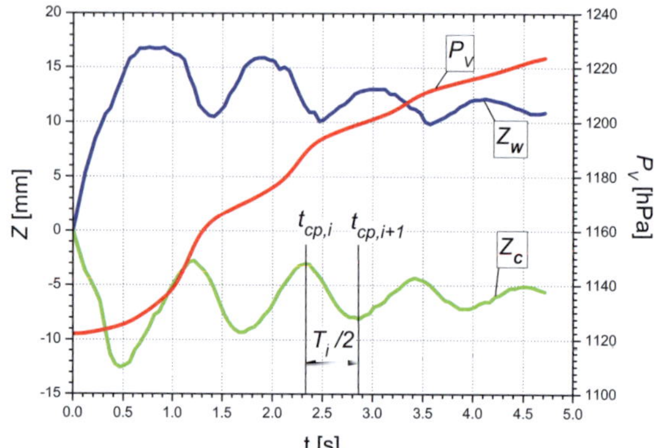

Fig. 5.3. Evolution of the deflections Z_w, Z_c and the pressure P_v for the experiment M17 with 1.9 K/mm. A definition is given of the period of the half oscillation $T_i/2$ between the times of two consecutive extrema $t_{cp,i}$ and $t_{cp,i+1}$ of the motion of Z_c.

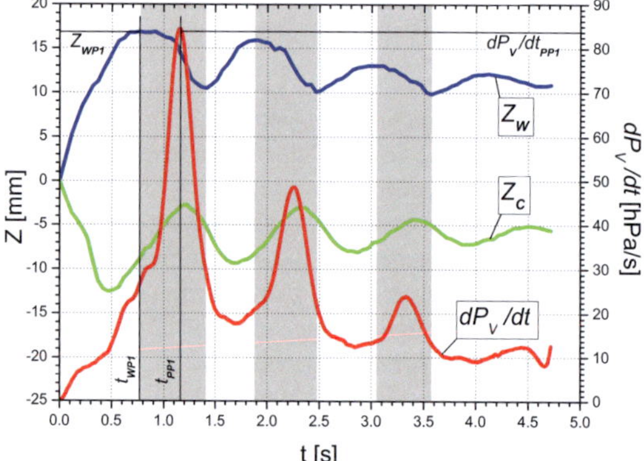

Fig. 5.4. Evolution of the deflections Z_w, Z_c and the pressure rate dP_v/dt for the experiment M17 with 1.9 K/mm. Gray areas mark the receding motion of the contact line. Definitions are given of the maximal deflection Z_{wp1} of the contact line at the time t_{wp1} and of the maximal peak dP_v/dt_{pp1} of the pressure rate at the time t_{pp1}.

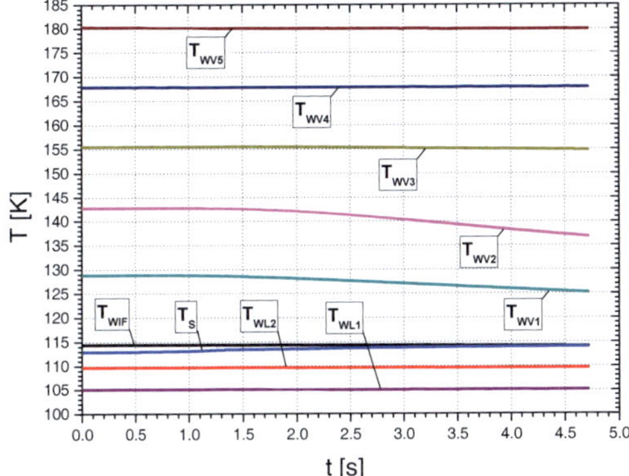

Fig. 5.5. Evolution of the wall temperature at the sensor positions in Fig. 4.4 and the saturation temperature T_S for the experiment M17 with 1.9 K/mm. It must be pointed out that the average meniscus position at the wall coincides with the position of the sensor T_{WV2}, situated on the outer side of the experiment cylinder.

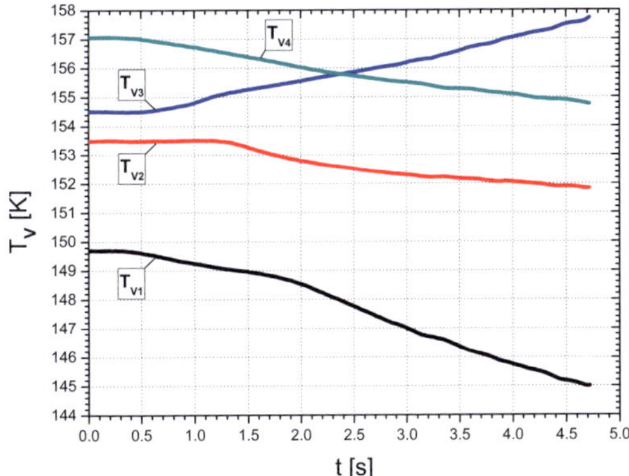

Fig. 5.6. Evolution of the vapor temperature at the sensor positions in Fig. 4.4 for the experiment M17 with 1.9 K/mm.

5.2 Interface dynamics

The dependence of the interface dynamics on the variation of the interfacial mass flux in the boundary conditions is presented in the following section through $\mathbf{E}_{cl}\mathbf{Oh}$, $\mathbf{EOh}$ and $\mathbf{Ev}_v$.

5.2.1 Center point

Frequency

As already mentioned the increase of $\Delta T_w/\Delta Z$ leads to an increase of the oscillation frequency of the center point and the appearance of the additional extrema. The number of the observed extrema for LAr, LCH$_4$, LH$_2$ and LNe is given in Tab. 5.1. The number is different for the different liquids and it increases with the increase of $\Delta T_w/\Delta Z$. I can define a mean frequency of the damped oscillation of the center

Table 5.1. Number of the observed extrema of the oscillation of the center point in dependence of $\Delta T_w/\Delta Z$.

$\Delta T_w/\Delta Z$ [K/mm]	0.15	0.73	1.34	1.4	2.0	2.5
N_{LAr}	3	3	4	4	4	5
$\Delta T_w/\Delta Z$ [K/mm]	0.2	0.7	1.3	1.9	2.5	2.9
N_{LCH_4}	6	6	7	8	8	9
$\Delta T_w/\Delta Z$ [K/mm]	0	0.3				
N_{LH_2}	5	6				
$\Delta T_w/\Delta Z$ [K/mm]	0	0.2	0.8			
N_{LNe}	2	2	2			

point after its first deflection as

$$\omega_{d,m} = \frac{\pi}{N-1} \sum_{i=1}^{N-1} \frac{1}{T_i/2} \tag{5.1}$$

where N is the corresponding value from Tab. 5.1 and $T_i/2$ is the period of the half cycle between the $(i+1)^{th}$ and the ith extremal value of the center point oscillation.

From Eq.(3.11) and Eq.(3.85) can be inferred that the undamped frequency of the center point would increase with the increase of wall superheat if the reorientation process is idealized as a step response of a linear second order system and an increase of the contact angle due to the superheat is assumed. Accordingly I present here the relation of the dimensionless oscillation frequency from the experiments to the wall superheat through $\mathbf{E}_{cl}\mathbf{Oh}$.

The definition of the dimensionless damped oscillation frequency given by [50] was presented in Eq. (2.37). One can use it in the form

$$\Omega_{d,m} = \omega_{d,m} t_{pu} \tag{5.2}$$

where $\omega_{d,m}$ is the mean damped frequency after the maximum deflection of the center point as defined in Eq. (5.1). In this manner one can investigate the influence of the variation of the specific boundary condition through the corresponding dimensionless numbers.

Thus, the frequency Ω_d exhibits a clear dependence on $\mathbf{E}_{cl}\mathbf{Oh}$ - it increase with the increase of $\mathbf{E}_{cl}\mathbf{Oh}$ and thereby with the increase of the evaporation from the contact line. This shows the dependence of the frequency on the contact line boundary condition. This dependence is shown in Fig. (5.7) for the experiments with liquid argon, methane, neon and hydrogen. The dependence seems to be approximately linear concerning the data points with argon and methane. The extrapolated value of the frequency for $\mathbf{E}_{cl}\mathbf{Oh} = 0$ is $\Omega_{d,m} \approx 3.0$. The value for the isothermal experiment (H2) with LH_2 is also $\Omega_{d,m} \approx 3$. So for these three liquids the extrapolated isothermal frequency agrees with the value $\Omega_{d,m} \geq 3$ as from Eq. (2.42). In the case of neon one has here $\Omega_{d,m} \approx 2.4$ which is lower than the expected minimal value of 3.

Next to the dependence on the contact line boundary condition the dependence of $\Omega_{d,m}$ on the mass balance boundary condition on interface through $\mathbf{E}\,\mathbf{Oh}$ can be investigated. This dependence is depicted in Fig. (5.8). The result is not as unambiguous as with $\mathbf{E}_{cl}\,\mathbf{Oh}$. Argon and methane show a clear trend, an increase of frequency with the increase of $\mathbf{E}\,\mathbf{Oh}$ and so with the evaporation from the interface. However, the results with argon differ with the two geometries. The data points are also more scattered.

In contrast to argon and methane the results with neon and hydrogen do not show the above mentioned trend. Besides this the extrapolated values to the isothermal condition are slightly different compared to the dependence on the evaporation from the contact line.

At last the dependence of the frequency on $\mathbf{Ev}_v$ is considered. This is the only dimensionless number on which the dimensionless mass flux at the interface j depends as given by Eq. (3.74). Due the fact the two term on the right hand side in Eq. (3.74) have opposite signs and that $\mathbf{Ev}_v$ appears in the denominator of the one term an increase of $\mathbf{Ev}_v$ would mean an increase of the absolute value of j. So in this case $\mathbf{Ev}_v$ is associated with the increase of the evaporation at the interface.

The dependence on $\mathbf{Ev}_v$ is depicted in Fig. (5.9). An evident increase of the frequency with the increase of $\mathbf{Ev}_v$ is present. So again can be inferred that the evaporation from the interface is related to the increase of oscillation frequency.

In summary it can be said that increase of evaporation, both from the interface and from the contact line, is related to the increase of the oscillation of the interface

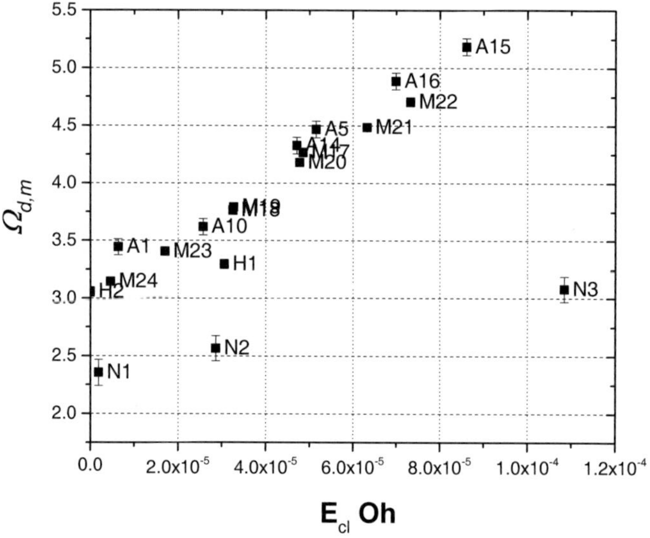

$\Omega_{d,m}$

$\mathbf{E}_{cl}\,\mathbf{Oh}$

Fig. 5.7. Dependence of the dimensionless mean oscillation frequency Ω_d of the center point on $\mathbf{E}_{cl}\,\mathbf{Oh}$.

oscillations. At least for argon and methane the extrapolated value to isothermal conditions seems to confirm the minimal value of 3.0 as proposed by [50].

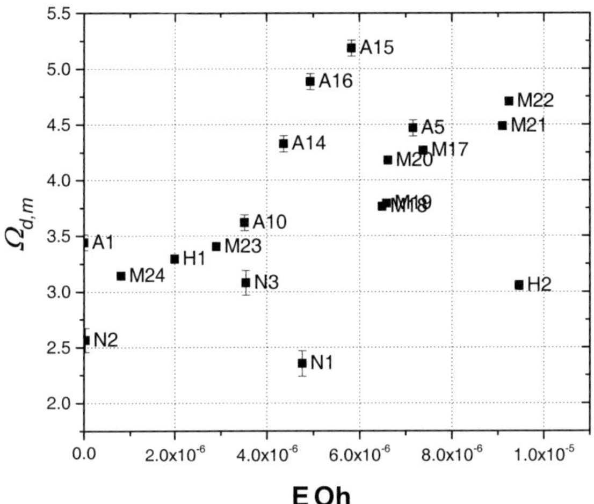

Fig. 5.8. Dependence of the dimensionless mean oscillation frequency Ω_d of the center point on **E Oh**.

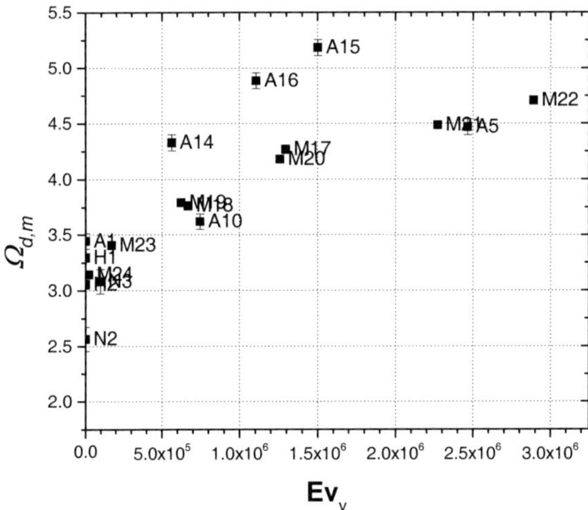

Fig. 5.9. Dependence of the dimensionless mean oscillation frequency Ω_d of the center point on **Ev$_v$**.

5.2.2 Contact line

Maximal deflection

The maximum deflection is a transient characteristic of the step response, which is proportional to the steady-state value according to Eq. (3.91). The steady-state value of the contact line on its turn is a monotonous function of the contact angle given by Eq. (3.12) - any increase of the contact angles results in a decrease of the steady-state deflection. This in turn would decrease the maximum deflection too. In the presence of evaporation at the contact line the contact angle is increased, Eq. (2.23).

In this context I want to show the relation between the dimensionless maximum deflection of the contact line Z_{wp1}/R from the experiments on $\mathbf{E}_{cl}\mathbf{Oh}$. The dependence is shown in Fig. (5.10). Data points only for liquid argon and methane are available. An unambiguous decrease of Z_{wp1}/R with a increase of $\mathbf{E}_{cl}\mathbf{Oh}$ is seen. The trend seems close to linear.

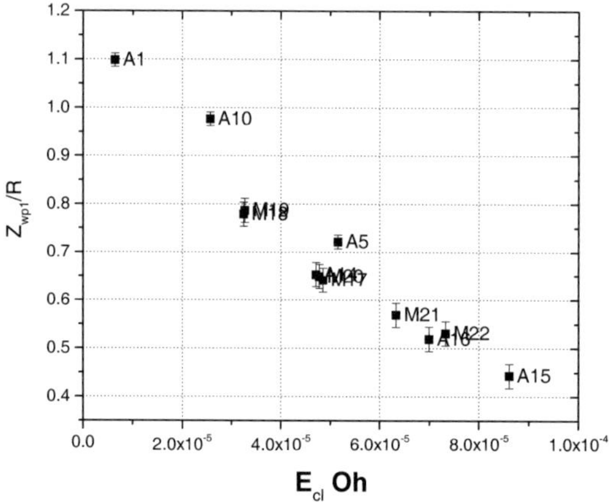

Fig. 5.10. Dependence of the dimensionless maximum deflection Z_{wp1}/R on $\mathbf{E}_{cl}\mathbf{Oh}$.

Since the linear step response of second order is only an approximation I use I investigate the dependence of the maximum defection of the contact line on the evaporation from the interface too. Accordingly the dependence of Z_{wp1}/R on $\mathbf{E}\,\mathbf{Oh}$ and $\mathbf{Ev}_v$ is shown in Fig. (5.11) and Fig. (5.12). In both cases a decrease is observed. However there is an apparent offset between the data points from the two geometries with argon.

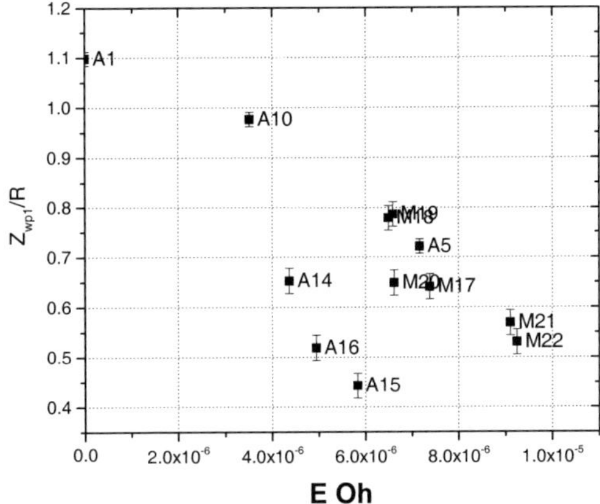

Fig. 5.11. Dependence of the dimensionless maximum deflection Z_{wp1}/R on **EOh**.

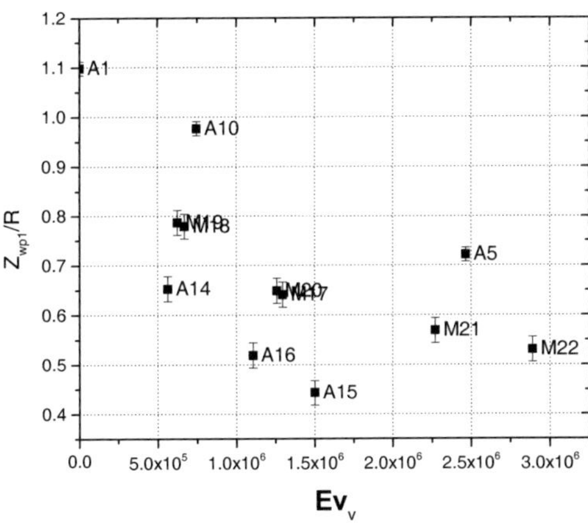

Fig. 5.12. Dependence of the dimensionless maximum deflection Z_{wp1}/R on **Ev$_v$**.

Velocity

Here I present the results with the contact line velocity. I restrict myself only to the experiments with LAr and LCH$_4$ because one can analyze only the mean velocity of the contact line for which based on [50] the observation of the peak value Z_{wp1} is needed. Only by these experiments could be observed Z_{wp1}. On the other hand the maximum contact line velocity was not accessible for observation.

Rise velocity

I have assumed the reorientation as a capillary-driven flow with phase change. Accordingly I use $U = u_{puR}$ from Eq. (3.34) for the non-dimensionalization of the mean rise velocity, defined by [50] as Z_{wp1}/t_{wp1}, by the relations

$$u_{r,m} = \frac{Z_{wp1}/t_{wp1}}{U} \tag{5.3}$$

where t_{wp1} is the peak time corresponding to Z_{wp1} as defined in Fig. (5.4).

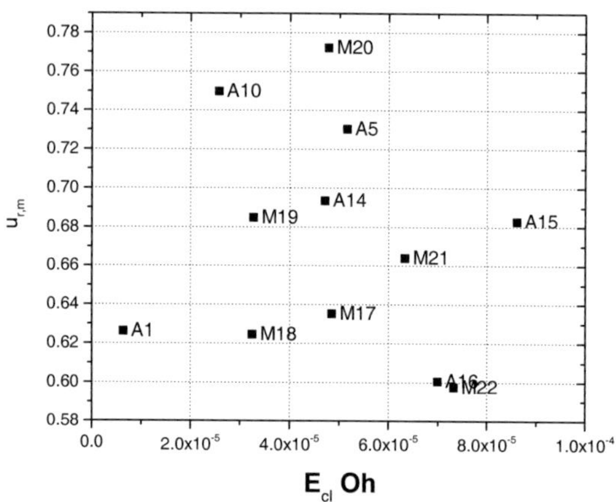

Fig. 5.13. Dependence of the dimensionless rise velocity $u_{r,m}$ of the contact line on $\mathbf{E}_{cl}\mathbf{Oh}$.

It must be said that no clear dependence of $u_{r,m}$ on any of the three dimensionless numbers could be observed, as it can be seen in Fig. (5.13-5.15). The data points are scattered irregularly between 0.6 and 0.8. This complies with the fact that the dimensional rise velocity does not show any clear trend with the increase of the of $\Delta T_w/\Delta Z$ either. So neither evaporation from the contact line (no dependence on $\mathbf{E}_{cl}\mathbf{Oh}$) nor the evaporation from the interface (no dependence on $\mathbf{EOh}$ and $\mathbf{Ev}_v$)

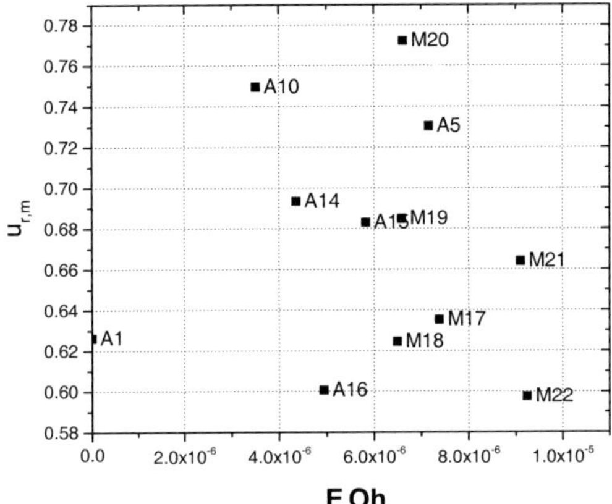

Fig. 5.14. Dependence of the dimensionless rise velocity $u_{r,m}$ of the contact line on **EOh**.

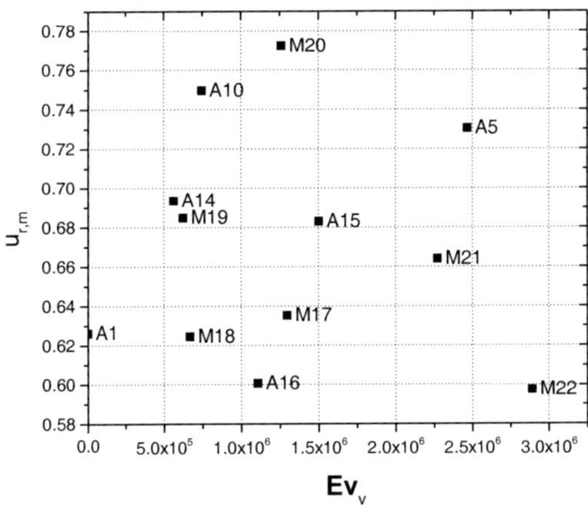

Fig. 5.15. Dependence of the dimensionless rise velocity $u_{r,m}$ of the contact line on **Ev$_v$**.

seem to affect the rise velocity of the contact line. Which makes me to conclude that the contact line rise is practically purely capillary-driven. This contrasts the suggestion by [50] that the evaporation at the contact line would affect its initial rise.

Velocity after Z_{wp1}

The velocity of the contact line after its maximum deflection Z_{wp1} was not investigated by [50]. The variation of the superheat in the contact line boundary condition should lead to a change of the contact line behavior and a variation of its velocity, which can be seen with argon and geometry A (A1, A10 and A5) - from a motionless contact line through a slow receding motion to a fast oscillating motion. Besides the influence of the evaporation from the interface could have an additional effect on the contact line motion.

Here I employ the above non-dimensionalization again once more with U. So I discuss here the dimensionless mean velocity u_{wom} in an analogous manner to the discussion of the rise velocity.

$$u_{wom} = \frac{1}{n-1} \sum_{i=1}^{n-1} u_{wom,i} \tag{5.4}$$

where n is the number of extrema of the contact line and $u_{wom,i}$ is given by

$$u_{wom,i} = \frac{\left| Z_{wp,i+1} - Z_{wp,i} \right| / \left(t_{wp,i+1} - t_{wp,i} \right)}{U} \tag{5.5}$$

I must point out that the contact line velocity after Z_{wp1} differs significantly from the rise velocity - even its highest values are around 2.5 smaller than the rise velocity. Besides this the dimensional mean velocity after the maximum deflection shows a clear increasing trend with the increase of $\Delta T_w / \Delta Z$, in contrast to the dimensional rise velocity.

The dependence of the dimensionless mean velocity u_{wom} is shown in Fig. (5.16 - Fig. 5.18). A clear increase of u_{wom} with increase of $\mathbf{E}_{cl}\mathbf{Oh}$, $\mathbf{E}\mathbf{Oh}$ and $\mathbf{Ev}_v$ is observed. So in contrast to the rise velocity the influence of evaporation from the contact line and the interface is apparent.

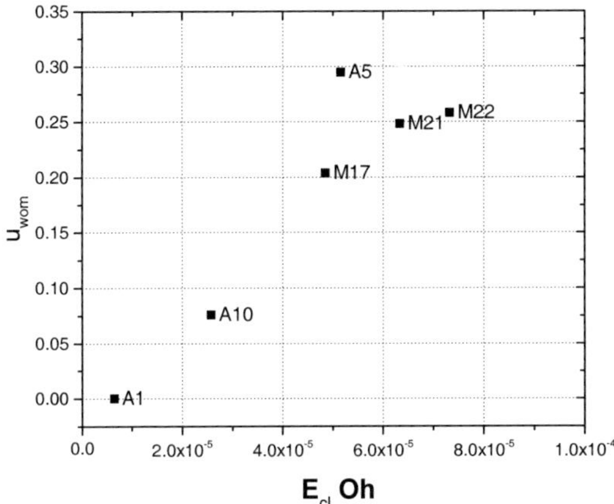

Fig. 5.16. Dependence of the dimensionless mean velocity u_{wom} of the contact line on $\mathbf{E}_{cl}\,\mathbf{Oh}$.

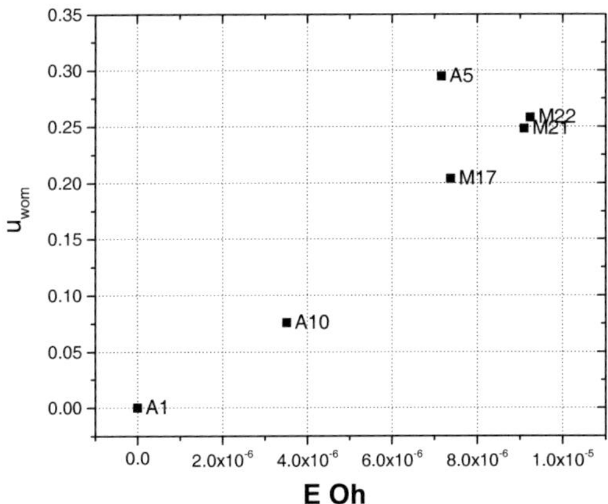

Fig. 5.17. Dependence of the dimensionless mean velocity u_{wom} of the contact line on $\mathbf{E}\,\mathbf{Oh}$.

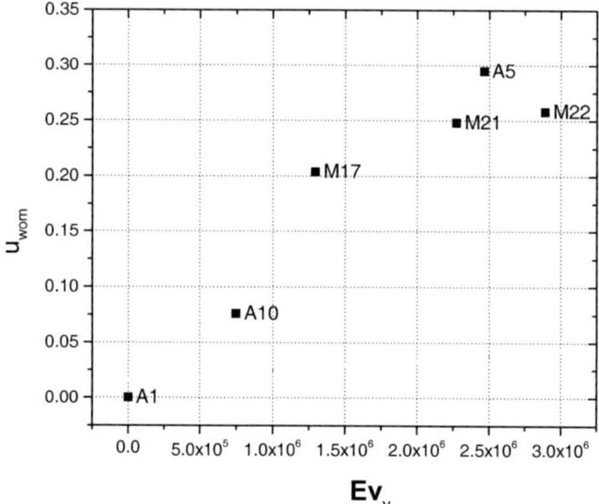

Fig. 5.18. Dependence of the dimensionless mean velocity u_{wom} of the contact line on $\mathbf{Ev}_v$.

6 Conclusions and outlook

The present work had two major goals concerning of the interface behavior during the reorientation of single species, two-phase, cryogenic liquids in partially filled containers under diabatic wall boundary conditions.

The first one was the development of a description of the reorientation expressing the influence of the evaporation from the interface and the contact line. For that purpose a mathematical description of the problem of the capillary driven flow in the presence of interfacial phase change and a superheated wall was formulated. Accordingly the treatment used by [50] and [29] was extended to the case of two-phase flow of cryogenic liquids with diabatic walls with a corresponding boundary condition at the contact line.

The second goal was the quantification of the influence of evaporation on the reorientation characteristics by executing drop tower experiments. For that purpose a variation of the liquids and of wall superheat was performed. Apart from this the presented experimental results extend the experimental results by [50] and numerical results by [29] to the range of dimensionless numbers for cryogenic liquids.

Regarding the first goal a non-dimensional form of the field equations and boundary conditions was developed. This included a boundary condition in general form and in specific form for cryogenic liquids. This enabled the presentation of the variation of the reorientation characteristics as functions of the variation of the dimensionless numbers from the the mass balance interface boundary condition and the contact line boundary condition, due to the variation of the interfacial mass flux.

Based on the assumption of the reorientation as a capillary flow with interfacial phase change it is possible to compare to a given extent (mainly qualitative) the results of this work to the literature. The dynamics and the steady-state of these examples in the literature vary with the variation of the interfacial mass flux.

It was observed that the dynamics of the reorientation vary also in this way. The oscillation frequency of the center point and the velocity of the contact line after the initial rise increase with the increase of evaporation. This was shown through the dependencies on $\mathbf{E}_{cl}\,\mathbf{Oh}$, $\mathbf{E}\,\mathbf{Oh}$ and $\mathbf{E}\mathbf{v}_v$. This is in accordance with the results by [64] and [61] who reported an increase of the interface oscillations by the capillary rise with evaporation.

On the other side there is a decrease of the maximal deflection of the contact line with the increase of evaporation. This is in accordance with the results from the spreading of evaporating droplets by [2] and the capillary rise with evaporation [64] and [61].

The extrapolation of the oscillation frequency of the center point from the present results to the value under isothermal conditions seems to be in accordance with the values proposed by [50] and [29] for argon, methane and hydrogen. For these substances the extrapolated value is around 3. Liquid neon shows a value of around 2.4.

The initial rise of the contact line, however, does not seem to be affected by the evaporation as suggested by [50].

As an outlook to future investigations of the reorientation I propose the inclusion of the vapor recoil effect in the boundary conditions at the interface, as in [11] and [2], and at the contact line. As for the oscillation description a non-linear step response would be more appropriate that my linear model.

References

1. Ajaev, V.: Interfacial Fluid Mechanics: A Mathematical Modeling Approach. Springer (2012)
2. Anderson, D.M., Davis, S.H.: The spreading of volatile liquid droplets on heated surfaces. Phys. Fluids **7**(2), 248–265 (1995)
3. Bänsch, E., Basting, S., Krahl, R.: Numerical simulation of two-phase flows with heat and mass transfer. Discrete and Continuous Dynamical Systems **35**(6), 2325–2347 (2015). DOI 10.3934/dcds.2015.35.2325
4. Bauer, H.F., Eidel, W.: Linear liquid oscillations in cylindrical container under zero-gravity. Appl. Microgravity Tech. **4**, 212–220 (1990)
5. Berg, J.C.: Wettability, *Surfactant Science Series*, vol. 49. Marcel Dekker, Inc., New York, Basel, Hong Kong (1993)
6. Blake, T.D.: The physics of moving wetting lines. J. Colloid Interf. Sci. **299**(1), 1–13 (2006)
7. Blake, T.D., Haynes, J.M.: Kinetics of liquid/ liquid displacement. J. Colloid Interf. Sci. **30**(3), 421–423 (1969)
8. Blake, T.D., Ruschak, K.J.: Wetting: Static and Dynamic Contact Lines, pp. 63–97. Springer Netherlands, Dordrecht (1997)
9. Bracke, M., de Voeght, F., Joos, P.: The kinetics of wetting: The dynamic contact angle. Prog. Coll. Pol. Sci. S. **79**, 142–149 (1989)
10. Brezesinski, G., Mögel, H.J.: Grenzflächen und Kolloide; Physicalisch-chemische Grundlagen. Spektrum Akademischer Verlag, Heidelberg, Berlin, Oxford. (1993). PHY 96
11. Burelbach, J.P., Bankoff, S.G., Davis, S.H.: Nonlinear stability of evaporating/condensing liquid films. J. Fluid Mech. **195**, 463–494 (1988)
12. Carey, V.P.: Liquid-Vapor Phase-Change Phenomena, an introduction to the thermophysics of vaporization and condensation processes in heat transfer equipment edn. No. ISBN: 0-89116-836-2 in Series in Chemical and Mechanical Engineering. Washington, D.C. : Hemisphere Pub. Corp., (1992). THE 20
13. Cazabat, A.M., Guena, G.: Evaporation of macroscopic sessile droplets. Soft Matter **6**, 2591–2612 (2010). DOI 10.1039/B924477H
14. Chao, L., Kamotani, Y., Ostrach, S.: G-jitter effects on the capillary surface motion in an open container under weightless condition. In: D.A. Siginer, M.M. Weislogel (eds.) Fluid Mechanics Phenomena in Microgravity, vol. AMD-Vol. 154, pp. 133–143. ASME, New York (1992)
15. Das, S.P., Hopfinger, E.J.: Mass transfer enhancement by gravity waves at a liquid-vapour interface. Int. J. Heat Mass Tran. **52**(5-6), 1400–1411 (2009)
16. Davis, S.H.: Moving contact lines and rivulet instabilities. part 1. the static rivulet. J. Fluid Mech. **98**(2), 225–242 (1980)
17. Davis, S.H.: Contact-line problems in fluid mechanics. J. Appl. Mech. **50**, 977–982 (1983)
18. Deen, W.: Analysis of Transport Phenomena. Topics in Chemical Engineering. OUP USA (1998)
19. Dreyer, M.E., Gerstmann, J., Stange, M., Rosendahl, U., Woelk, G., Rath, H.J.: Capillary effects under low gravity, part i: Surface settling, capillary rise and critical velocities. Space Forum **3**, 87–136 (1998)
20. Dussan, E.B.: The moving contact line: The slip boundary condition. J. Fluid Mech. **77**(4), 665–684 (1976)

21. Dussan, E.B.: On the spreading of liquids on solid surfaces: Static and dynamic contact lines. Annu. Rev. Fluid Mech. **11**, 371–400 (1979)
22. Eckels, P., Stewart, R.: CryoComp. Thermal Analysis Software Version 5.1, Eckels Engineering Inc (2010)
23. Eggers, J., Stone, H.A.: Characteristic lengths at moving contact lines for a perfectly wetting fluid: the influence of speed on the dynamic contact angle. Journal of Fluid Mechanics **505**, 309–321 (2004). DOI 10.1017/S0022112004008663
24. Ehrhard, P.: Experiments of isothermal and non-isothermal spreading. J. Fluid Mech. **257**, 463–483 (1993)
25. Ehrhard, P., Davis, S.H.: Non-isothermal spreading of liquid drops on horizontal plates. J. Fluid Mech. **229**, 365–388 (1991)
26. Fermigier, M., Jenffer, P.: An experimental investigation of the dynamic contact angle in liquid-liquid systems. Journal of Colloid and Interface Science **146**(1), 226 – 241 (1991). DOI http://dx.doi.org/10.1016/0021-9797(91)90020-9
27. Friz, G.: Über den dynamischen Randwinkel im Fall der vollständigen Benetzung. Z. angew. Phys. **19.4**, 374–378 (1965)
28. de Gennes, P.G.: Wetting: Statics and dynamics. Rev. Mod. Phys. **57**(3), 827–863 (1985)
29. Gerstmann, J.: Numerische Untersuchungen zur Schwingung freier Fluessigkeitsoberflaechen, *Stroemungstechnik*, vol. 464. VDI Verlag, Duesseldorf (2004)
30. Gerstmann, J., Dreyer, M.E.: Axisymmetric surface oscillations in a cylindrical container with compensated gravity. Ann. NY Acad. Sci. **1077**, 328–350 (2006)
31. Gerstmann, J., Michaelis, M., Dreyer, M.: Capillary driven oscillations of a free liquid interface under non-isothermal conditions. In: PAMM, vol. 4, pp. 436–437 (2004)
32. Greenspan, H.P.: On the motion of a small viscous droplet that wets a surface. J. Fluid Mech. **84**(1), 125–143 (1978)
33. Haley, P.J., Miskis, M.J.: The effect of the contact line on droplet spreading. J. Fluid Mech. **223**, 57–81 (1991)
34. Hirt, C.W., Nichols, B.D.: Volume of fluid (vof) method for the dynamics of free boundaries. J. Comput. Phys. **39**(1), 201–225 (1981)
35. Hocking, L.M.: On contact angles in evaporating liquids. Physics of Fluids **7**(12), 2950–2955 (1995)
36. Hoffman, R.L.: A study of the advancing interface: I. interface shape in liquid-gas systems. J. Colloid Interf. Sci. **50**(2), 228–241 (1975)
37. Hoffman, R.L.: A study of the advancing interface. ii. theoretical prediction of the dynamic contact angle in liquid-gas systems. J. Colloid Interf. Sci. **94**(2), 470–486 (1983)
38. Ibrahim, R.A.: Liquid Sloshing Dynamics. ISBN: 978-0-521-83885-6. Cambridge University Press (2005)
39. Jiang, T.S., Oh, S.G., Slattery, J.C.: Correlation for dynamic contact angle. J. Colloid Interf. Sci. **69**(1), 74–77 (1979)
40. Kamotani, Y., Chao, L., Ostrach, S., Zhang, H.: Effects of g jitter on free-surface motion in a cavity. J. Spacecraft Rockets **32**(1), 177–183 (1995)
41. Kennard, E.: Kinetic theory of gases: with an introduction to statistical mechanics. International series in pure and applied physics. McGraw-Hill (1938)
42. Kistler, S.F.: Chapter 6. In: J.C. Berg (ed.) Hydrodynamics of Wetting, p. 311. Marcel Dekker, Inc., New York (1993)
43. Klein, R., Botta, N., Schneider, T., Munz, C., Roller, S., Meister, A., Hoffmann, L., Sonar, T.: Asymptotic adaptive methods for multi-scale problems in fluid mechanics. In: Practical Asymptotics, pp. 261–343. Springer (2001)
44. Krahl, R., Bänsch, E.: On the stability of an evaporating liquid surface. Fluid Dynamics Research **44**(031409), 1–16 (2012)

45. Kulev, N., Basting, S., Baensch, E., Dreyer, M.: Interface reorientation of cryogenic liquids under non-isothermal boundary conditions. Cryogenics **62**, 48–59 (2014)

46. Kulev, N., Dreyer, M.E.: Drop tower experiments on non-isothermal reorientation of cryogenic liquids. Microgravity Sci. Technol. **22**(4), 463–474 (2010)

47. Lamb, H.: Hydrodynamics. The University Press (1932)

48. Landau, L.D., Lifschitz, E.M.: Lehrbuch der Theoretischen Physik, Band VI, Hydrodynamik. Akademie Verlag, Berlin (1991)

49. Lemmon, E.W., Peskin, A.P., McLinden, M.O., Friend, D.: NIST Thermodynamic and transport properties of pure fluids. NIST Standard Reference Database 12 Version 5.0, U. S. Department of Commerce (2000)

50. Michaelis, M.: Kapillarinduzierte Schwingungen freier Fluessigkeitsoberflaechen, *Stroemungstechnik*, vol. 454. VDI Verlag, Duesseldorf (2003)

51. Michaelis, M., Dreyer, M.E.: Test-case no 31: Reorientation of a free liquid interface in a partly filled right circular cylinder upon gravity step reduction. Multiphase Science and Technology **16**(1-3), 219–238 (2004)

52. Michaelis, M., Dreyer, M.E., Rath, H.J.: Experimental investigation of the liquid interface reorientation upon step reduction in gravity. Ann. NY Acad. Sci. **974**, 246–260 (2002)

53. Michaelis, M., Gerstmann, J., Dreyer, M.E., Rath, H.J.: Damping behaviour of the free liquid interface oscillation upon step reduction in gravity. In: Proc. Appl. Math. Mech., vol. 2, pp. 320–321 (2003)

54. Morris, S.J.S.: Contact angles for evaporating liquids predicted and compared with existing experiments. J. Fluid Mech. **432**, 1–30 (2001)

55. van Mourik, S., Veldman, A.E.P., Dreyer, M.E.: Simulation of capillary flow with a dynamic contact angle. Microgravity Sci. Technol. **17**(3), 87–94 (2005)

56. Mourik, S.V.: Numerical modelling of the dynamic contact angle. Master's thesis, University of Groningen (2002)

57. Myers, D.: Surfaces, interfaces, and colloids: principles and applications. Wiley-VCH (1999)

58. Ogata, K.: System Dynamics. Pearson/Prentice Hall (2004)

59. Ogata, K.: Modern Control Engineering. Instrumentation and controls series. Prentice Hall (2010)

60. Oron, A., Davis, S.H., Bankoff, G.S.: Long-scale evolution of thin liquid films. Rev. Mod. Phys. **69**(3), 931–980 (1997)

61. Polansky, J., Kaya, T.: An experimental and numerical study of capillary rise with evaporation. Int. J. Therm. Sci. **91**, 25–33 (2015)

62. Potash, M.J., Wayner, P.C.J.: Evaporation from a two-dimensional extended meniscus. Int. J. Heat Mass Tran. **15**, 1851–1863 (1972)

63. Raj, R., Kunkelmann, C., P. Stephan, P., Plawsky, J., Kim, J.: Contact line behavior for a highly wetting fluid under superheated conditions. Int. J. Heat Mass Tran. **55**(9-10), 2664–2675 (2012)

64. Ramon, G., Oron, A.: Capillary rise of a meniscus with phase change. J. Colloid Interf. Sci. **327**(1), 145–151 (2008)

65. Rand, R.H.: The dynamics of an evaporating meniscus. Acta Mech. **39**(1-4), 135–146 (1978)

66. Sánchez, S., Méndez, F., Bautista, O.: Capillary rise in a circular tube with interfacial condensation process. International Journal of Thermal Sciences **50**(12), 2422 – 2429 (2011)

67. Satterlee, H., Reynolds, W.: Dynamics of the Free Liquid Surface in Cylindrical Containers Under Strong Capillary and Weak Gravity Conditions. Stanford University. Dept. of Mechanical Engineering. Thermosciences Division. National Science Foundation (U.S.) (1964)

68. Schwartz, A.M., Tejada, S.B.: Studies of dynamic contact angles on solids. J. Colloid Interf. Sci. **38**(2), 359–375 (1972)

69. Seebergh, J.E., Berg, J.C.: Dynamic wetting in the low capillary number regime. Chem. Eng. Sci. **47**(17-18), 4455–4464 (1992)

70. Shikhmurzaev, Y.D.: Capillary Flows with Forming Interfaces. ISBN: 978-1-58488-748-5. Chapman and Hall/CRC, Taylor and Francis Group (2008)

71. Siegert, C.E., Otto, E.W., Petrash, D.A.: Behavior of liquid-vapor interface of cryogenic liquids during weightlessness. NASA Technical Report, NASA LeRC (1965)

72. Siegert, C.E., Petrash, D.A., Otto, E.W.: Time response of liquid-vapor interface after entering weightlessness. Tech. Rep. NASA TN D-2458, Lewis Research Center, Cleveland, Ohio (1964)

73. Sodtke, C., Ajaev, V.S., Stephan, P.: Dynamics of volatile liquid droplets on heated surfaces: theory versus experiment. J. Fluid Mech. **610**, 343–362 (2008)

74. Stephan, P.C., Busse, C.A.: Analysis of the heat transfer coefficient of grooved heat pipe evaporator walls. Int. J. Heat Mass Tran. **35**(2), 383–391 (1992)

75. Stief, M., Gerstmann, J., Dreyer, M.E.: Reorientation of cryogenic fluids upon step reduction of gravity. In: PAMM, vol. 5, pp. 553–554. Wiley-Interscience (2005)

76. Ström, G., Fredriksson, M., Stenius, P., Radoev, B.: Kinetics of steady-state wetting. Journal of Colloid and Interface Science **134**(1), 107 – 116 (1990)

77. Tanner, L.H.: The spreading of silicone oil drops on horizontal surfaces. J. Phys. D Appl. Phys. **12**(9), 1473–1484 (1979)

78. Voinov, O.V.: Hydrodynamics of wetting. Fluid Dyn. **11**(5), 714–721 (1976)

79. Wayner, P.C.: The effect of interfacial mass transport on flow in thin liquid films. Colloids and Surfaces **52**, 71 – 84 (1991). DOI http://dx.doi.org/10.1016/0166-6622(91)80006-A

80. Wayner, P.C.J., Coccio, C.L.: Heat and mass transfer in the vicinity of the triple interline of a meniscus. AICHE J. **17**(3), 569–574 (1971)

81. Weinstein, S.J., Palmer, H.J.: Capillary hydrodynamics and interfacial phenomena. In: S.F. Kistler, P.M. Schweizer (eds.) Liquid Film Coating: Scientific principles and their technological implications, chap. 2, pp. 19–62. Springer Netherlands, Dordrecht (1997)

82. Weislogel, M.M., Ross, H.D.: Surface reorientation and settling in cylinders upon step reduction in gravity. Microgravity Sci. Technol. **3**(1), 24–32 (1990)

83. Woelk, G., Dreyer, M.E., Rath, H.J., Weisvogel, M.M.: Damped oscillations of a liquid/gas surface upon step reduction in gravity. J. Spacecraft Rockets **34**(1), 110–117 (1997)

A Appendix

A.1 Experimental parameters

A.1.1 Initial conditions

Table A.1. Experiment conditions prior to the reduction of gravity with liquid argon with geometry A alongside with the corresponding heating times.

$\Delta T_w/\Delta Z$	P_v	T_S	T_{wf}	T_l	T_{wl}	T_v	T_{wv}	t_T	Experiment
[K/mm]	[hPa]	[K]	[K]	[K]	[K]	[K]	[K]	[s]	
0.15	934	86.5	88.0	86.6	87.3	95.8	98.8	1320.0	A1
0.73	1051	87.7	90.3	84.8	86.6	131.7	141.3	540.0	A10
1.34	1441	90.8	98.4	85.0	86.4	175.3	192.1	750.0	A5

Table A.2. Experiment conditions prior to the reduction of gravity with liquid argon with geometry B alongside with the corresponding heating times.

$\Delta T_w/\Delta Z$ [K/mm]	P_v [hPa]	T_S [K]	T_L [K]	T_{WIF} [K]	T_{WL1} [K]	T_{WL2} [K]	T_{WV1} [K]	T_{WV2} [K]	T_{WV3} [K]	T_{WV4} [K]	T_{WV5} [K]	T_{V1} [K]	T_{V2} [K]	T_{V3} [K]	T_{V4} [K]	t_T [s]	Experiment
1.4	1324	89.9	85.1	92.1	86.3	88.4	102.7	112.2	121.5	130.3	139.6	117.3	119.2	119.7	120.7	689.9	A14
2.0	1211	89.0	84.4	94.6	84.9	87.9	108.9	121.7	135.0	148.8	163.7	129.5	132.6	133.3	133.9	489.3	A16
2.5	1371	90.3	84.7	95.7	85.5	88.3	114.4	130.8	147.1	164.1	182.3	139.2	142.5	143.7	143.4	515.7	A15

Table A.3. Experiment conditions prior to the reduction of gravity with liquid methane with geometry B alongside with the corresponding heating times.

$\Delta T_w/\Delta Z$ [K/mm]	P_v [hPa]	T_S [K]	T_{WIF} [K]	T_{WL1} [K]	T_{WL2} [K]	T_{WV1} [K]	T_{WV2} [K]	T_{WV3} [K]	T_{WV4} [K]	T_{WV5} [K]	T_{V1} [K]	T_{V2} [K]	T_{V3} [K]	T_{V4} [K]	t_T [s]	Experiment
0.2	477	103.3	103.3	102.4	102.9	104.8	106.1	107.9	109.1	110.6	107.8	108.2	108.6	108.7	597.6	M24
0.7	662	106.7	107.1	103.5	105.5	112.1	117.0	122.1	126.5	131.0	120.2	121.7	122.2	122.5	951.4	M23
1.3	877	109.9	110.9	102.9	107.4	120.9	130.4	139.5	147.9	156.5	135.3	138.1	138.9	139.7	1120.1	M18
1.3	1045	112.0	112.7	105.0	109.5	122.3	131.5	140.2	148.2	156.7	136.3	139.2	140.0	140.9	1221.2	M19
1.9	1122	112.9	114.4	105.1	109.7	128.9	142.7	155.5	167.8	180.3	149.7	153.5	154.5	157.1	830.9	M17
1.9	926	110.6	112.1	103.5	107.5	126.4	140.1	153.1	165.9	180.1	147.9	152.1	153.5	154.9	723.4	M20
2.5	891	110.1	112.8	100.3	106.1	132.4	150.3	166.8	183.3	200.6	159.0	163.9	165.7	168.3	727.2	M21
2.9	838	109.4	112.7	99.4	105.1	135.2	155.7	175.0	194.4	215.3	165.4	172.4	174.9	176.5	619.3	M22

Table A.4. Experiment conditions prior to the reduction of gravity with liquid neon with geometry B alongside with the corresponding heating times.

$\Delta T_w/\Delta Z$ [K/mm]	P_v [hPa]	T_S [K]	T_{WIF} [K]	T_{WL1} [K]	T_{WL2} [K]	T_{WV1} [K]	T_{WV2} [K]	T_{WV3} [K]	T_{WV4} [K]	T_{WV5} [K]	T_{V1} [K]	T_{V2} [K]	T_{V3} [K]	T_{V4} [K]	t_T [s]	Experiment
0.0	1297	27.9	28.3	27.0	28.2	28.5	28.6	29.0	29.2	29.3	29.0	29.2	29.9	31.0	0	N1
0.2	962	26.9	27.7	26.9	27.2	29.8	31.6	33.6	34.9	35.3	32.7	34.4	35.1	35.3	230	N2
0.8	1121	27.4	29.0	26.7	27.3		42.4	47.3	50.8	53.8	45.3	49.6	50.3	51.2	700	N3

Table A.5. Experiment conditions prior to the reduction of gravity with liquid argon with geometry B alongside with the corresponding heating times.

$\Delta T_w/\Delta Z$ [K/mm]	P_v [hPa]	T_S [K]	T_{WIF} [K]	T_{WL1} [K]	T_{WL2} [K]	T_{WV1} [K]	T_{WV2} [K]	T_{WV3} [K]	T_{WV4} [K]	T_{WV5} [K]	T_{V1} [K]	T_{V2} [K]	T_{V3} [K]	T_{V4} [K]	t_T [s]	Exp.
0.0	1091	20.5	20.5	18.1	20.4	20.5	20.6	20.6	20.6	20.6	20.8	20.8	21.0	22.4	0	H2
0.3	757	19.3	20.0	18.8	19.5	22.2	24.4	26.5	28.1	29.3	25.7	28.1	30.1	34.8	0	H1
0.7	852	19.7	20.9	18.6	19.6	27.0	32.8	38.2	41.9	44.9	35.5	41.7	43.2	44.5	500	H3

A.1.2 Scales

Table A.6. Scales by the experiments with liquid argon with geometry A.

$\Delta T_w/\Delta Z$	u_{puR}	t_{puR}	δ_T	Θ	Θ_v	Θ_w	J	J_{cl}	Experiment
[K/mm]	[mm/s]	[s]	[m]	[K]	[K]	[K]	[10^{-4}kg s^{-1}]	[10^{-4}kg s^{-1}]	
0.15	18.6	1.41	0.0105	0.0	9.2	5.3	0	4.1	A1
0.73	18.5	1.42	0.0067	2.9	44.0	21.2	3.5	25.5	A10
1.34	18.0	1.46	0.0078	5.8	84.5	41.8	5.9	42.3	A5

Table A.7. Scales by the experiments with liquid methane with geometry B.

$\Delta T_w/\Delta Z$	u_{puR}	t_{puR}	δ_T	Θ	Θ_v	Θ_w	J	J_{cl}	Experiment
[K/mm]	[mm/s]	[s]	[m]	[K]	[K]	[K]	[10^{-4}kg s^{-1}]	[10^{-4}kg s^{-1}]	
0.2	36.8	0.71	0.0089	0.9	4.5	5.2	0.4	2.2	M24
0.7	35.9	0.73	0.0111	3.2	13.5	18.7	1.1	6.2	M23
1.3	35.1	0.75	0.0119	7.0	25.4	35.1	2.1	10.7	M18
1.3	34.6	0.76	0.0123	7.0	24.3	34.8	2.0	10.1	M19
1.9	34.3	0.76	0.0102	7.8	36.8	51.3	2.8	18.1	M17
1.9	34.9	0.75	0.0095	7.1	37.3	51.3	2.7	19.4	M20
2.5	35.1	0.75	0.0096	9.8	48.9	68.2	3.7	25.8	M21
2.9	35.3	0.74	0.0089	10.0	56.0	79.3	4.1	32.5	M22

Table A.8. Scales by the experiments with liquid argon with geometry B.

$\Delta T_w/\Delta Z$	u_{puR}	t_{puR}	δ_T	Θ	Θ_v	Θ_w	J	J_{cl}	Experiment
[K/mm]	[mm/s]	[s]	[m]	[K]	[K]	[K]	[10^{-4}kg s^{-1}]	[10^{-4}kg s^{-1}]	
1.4	18.1	1.45	0.00751	3.6	30.8	38.9	3.8	41.1	A14
2.0	18.3	1.43	0.00635	4.1	44.9	58.0	5.6	73.0	A16
2.5	18.1	1.45	0.00648	4.8	53.1	70.9	5.9	86.6	A15

Table A.9. Scales by the experiments with liquid neon with geometry B.

$\Delta T_w/\Delta Z$	u_{puR}	t_{puR}	δ_T	Θ	Θ_v	Θ_w	J	J_{cl}	Experiment
[K/mm]	[mm/s]	[s]	[m]	[K]	[K]	[K]	$[10^{-4}\,\mathrm{kg\,s^{-1}}]$	$[10^{-4}\,\mathrm{kg\,s^{-1}}]$	
0.0	11.6	2.26		1.0	3.1	0.4			N1
0.2	12.1	2.17	0.004	0.0	8.4	6.0	0.03	27.4	N2
0.8	11.8	2.22	0.00691	0.7	23.8	22.5	1.9	58.4	N3

Table A.10. Scales by the experiments with liquid hydrogen with geometry B.

$\Delta T_w/\Delta Z$	u_{puR}	t_{puR}	δ_T	Θ	Θ_v	Θ_w	J	J_{cl}	Experiment
[K/mm]	[mm/s]	[s]	[m]	[K]	[K]	[K]	$[10^{-4}\,\mathrm{kg\,s^{-1}}]$	$[10^{-4}\,\mathrm{kg\,s^{-1}}]$	
0.0	32.3	0.81		2.4	1.9	0			H2
0.3	33.6	0.78		0.6	15.5	8.5			H1
0.7	33.2	0.79	0.00879	1.1	24.8	19.5	0.3	5.1	H3

A.1.3 Dimensionless numbers

Table A.11. Dimensionless numbers by the experiments with liquid argon with geometry A.

$\Delta T_w/\Delta Z$	Ev	E	E_{cl}	Oh	E Oh	E_{cl} Oh	Experiment
[K/mm]	[10^6]	[10^{-2}]	[10^{-1}]	[10^{-4}]	[10^{-6}]	[10^{-5}]	
0.15	0.00	0.00	0.16	3.96	0.00	0.64	A1
0.73	0.75	0.91	0.67	3.87	3.52	2.57	A10
1.34	2.47	1.95	1.40	3.68	7.16	5.16	A5

Table A.12. Dimensionless numbers by the experiments with liquid methane with geometry B.

$\Delta T_w/\Delta Z$	Ev	E	E_{cl}	Oh	E Oh	E_{cl} Oh	Experiment
[K/mm]	[10^6]	[10^{-2}]	[10^{-1}]	[10^{-4}]	[10^{-6}]	[10^{-5}]	
0.2	0.02	0.23	0.14	3.41	0.80	0.47	M24
0.7	0.17	0.90	0.52	3.25	2.90	1.70	M23
1.3	0.67	2.09	1.04	3.11	6.49	3.25	M18
1.3	0.63	2.17	1.08	3.03	6.58	3.27	M19
1.9	1.30	2.46	1.62	3.00	7.38	4.85	M17
1.9	1.26	2.14	1.55	3.08	6.61	4.78	M20
2.5	2.27	2.93	2.04	3.10	9.10	6.33	M21
2.9	2.89	2.95	2.34	3.13	9.24	7.33	M22

Table A.13. Dimensionless numbers by the experiments with liquid argon with geometry B.

$\Delta T_w/\Delta Z$	Ev	E	E_{cl}	Oh	E Oh	E_{cl} Oh	Experiment
[K/mm]	[10^6]	[10^{-2}]	[10^{-1}]	[10^{-4}]	[10^{-6}]	[10^{-5}]	
1.4	0.56	1.19	1.28	3.68	4.36	4.71	A14
2.0	1.11	1.33	1.88	3.73	4.94	6.99	A16
2.5	1.50	1.60	2.36	3.65	5.83	8.61	A15

Table A.14. Dimensionless numbers by the experiments with liquid neon with geometry B.

$\Delta T_w/\Delta Z$	Ev	E	E_{cl}	Oh	E Oh	E_{cl} Oh	Experiment
[K/mm]	$[10^6]$	$[10^{-2}]$	$[10^{-1}]$	$[10^{-4}]$	$[10^{-6}]$	$[10^{-5}]$	
0.0		1.6	0.06	2.97	4.76	0.19	N1
0.2	0.0005	0.0	0.93	3.10	0.03	2.89	N2
0.8	0.097	1.17	3.58	3.03	3.55	10.86	N3

Table A.15. Dimensionless numbers by the experiments with liquid hydrogen with geometry B.

$\Delta T_w/\Delta Z$	Ev	E	E_{cl}	Oh	E Oh	E_{cl} Oh	Experiment
[K/mm]	$[10^6]$	$[10^{-2}]$	$[10^{-1}]$	$[10^{-4}]$	$[10^{-6}]$	$[10^{-5}]$	
0.0	0.0	4.32	0.0001	2.19	0.007	0.005	H2
0.3	0.0	0.88	1.34	2.28	13.4	8.5	H1
0.7	36.3	1.83	3.20	2.25	32.0	21.4	H3

A.2 Experimental results

A.2.1 Tables with the values of the interface characteristics

Table A.16. Values of the reorientation characteristics of the interface with liquid argon and geometry A.

$\Delta T_w/\Delta Z$ [K/mm]		0.15	0.73	1.34
Exp.	ID	A1	A10	A5
t_{cp1}	[s]	0.9	0.9	0.9
Z_{cp1}	[mm]	-18.0	-16.9	-16.3
t_{cp2}	[s]	2.6	2.6	2.2
Z_{cp2}	[mm]	-8.3	-5.8	-4.0
t_{cp3}	[s]	3.6	3.6	2.9
Z_{cp3}	[mm]	-12.2	-11.5	-11.7
t_{cp4}	[s]	-	-	4.0
Z_{cp4}	[mm]	-	-	-5.0
ω_d	[1/s]	2.45	2.55	3.07
t_{wp1}	[s]	2.4	1.8	1.4
Z_{wp1}	[mm]	28.0	24.9	18.4
t_{wp2}	[s]	-	3.8	2.4
Z_{wp2}	[mm]	-	22.1	9.6
t_{wp3}	[s]	-	-	3.4
Z_{wp3}	[mm]	-	-	15.5
t_{wp4}	[s]	-	-	4.1
Z_{wp4}	[mm]	-	-	10.0
u_{wr}	[mm/s]	11.67	13.83	13.14
u_{wp12}	[mm/s]		-1.4	-4.4
u_{wp23}	[mm/s]			5.9
u_{wp34}	[mm/s]			-5.5
$|u_{wom}|$	[mm/s]	0.0	1.4	5.3
$u_{wom,rec}$	[mm/s]	0.0	-1.4	-5.0

Table A.17. Values of the reorientation characteristics of the interface with liquid argon and geometry B.

$\Delta T_w / \Delta Z$ [K/mm]		1.4	2.0	2.5
Exp.	ID	A14	A16	A15
t_{cp1}	[s]	0.82	0.84	0.76
Z_{cp1}	[mm]	-13.7	-12.3	-12.8
t_{cp2}	[s]	2.0	1.84	1.66
Z_{cp2}	[mm]	-4.6	-4.4	-4.6
t_{cp3}	[s]	2.93	2.78	2.62
Z_{cp3}	[mm]	-9.8	-8.9	-9.6
t_{cp4}	[s]	4.00	3.62	3.56
Z_{cp4}	[mm]	-5.1	-5.1	-4.8
t_{cp5}	[s]		-	4.34
Z_{cp5}	[mm]		-	-7.9
ω_d	[1/s]	3.00	3.40	3.58
t_{wp1}	[s]	1.3	1.24	0.94
Z_{wp1}	[mm]	16.2	13.6	11.6
t_{wp2}	[s]	-		
Z_{wp2}	[mm]	-		
t_{wp3}	[s]	-	-	
Z_{wp3}	[mm]	-	-	
t_{wp4}	[s]	-	-	
Z_{wp4}	[mm]	-	-	
u_{wr}	[mm/s]	12.57	10.97	12.34

Table A.18. Values of the reorientation characteristics of the interface with liquid methane and geometry B.

$\Delta T_w/\Delta Z$ [K/mm]		0.2	0.7	1.3	1.3	1.9	1.9	2.5	2.9
Exp.	ID	M24	M23	M18	M19	M17	M20	M21	M22
t_{cp1}	[s]	0.48	0.48	0.50	0.48	0.47	0.46	0.45	0.42
Z_{cp1}	[mm]	-13.7	-13.6	-13.2	-13.7	-12.5	-13.7	-13.2	-13.4
t_{cp2}	[s]	1.4	1.38	1.25	1.35	1.22	1.16	1.09	1.04
Z_{cp2}	[mm]	-6.3	-5.0	-3.5	-4.5	-2.7	-4.3	-3.7	-3.7
t_{cp3}	[s]	1.98	1.88	1.80	1.86	1.69	1.68	1.54	1.52
Z_{cp3}	[mm]	-9.5	-10.0	-9.8	-10.2	-9.3	-10.7	-10.2	-10.6
t_{cp4}	[s]	2.73	2.73	2.53	2.55	2.34	2.33	2.14	2.06
Z_{cp4}	[mm]	-6.5	-5.1	-4.2	-4.3	-3.1	-4.7	-4.5	-4.4
t_{cp5}	[s]	3.36	3.3	3.1	3.15	2.87	2.81	2.63	2.48
Z_{cp5}	[mm]	-8.5	-9.1	-8.9	-9.2	-8.1	-9.2	-8.8	-8.8
t_{cp6}	[s]	4.23	4.02	3.78	3.84	3.42	3.43	3.17	3.03
Z_{cp6}	[mm]	-6.8	-6.1	-5.3	-5.4	-4.3	-5.8	-5.6	-5.7
t_{cp7}	[s]			4.33	4.41	3.92	3.91	3.65	3.55
Z_{cp7}	[mm]			-8.1	-8.7	-7.1	-8.5	-7.5	-7.7
t_{cp8}	[s]					4.49	4.48	4.22	4.00
Z_{cp8}	[mm]					-5.1	-6.3	-6.1	-6.1
t_{cp9}	[s]								4.45
Z_{cp9}	[mm]								-7.2
ω_d	[1/s]	4.42	4.67	5.04	5.00	5.59	5.58	6.00	6.33
t_{wp1}	[s]			0.93	0.87	0.77	0.63	0.64	0.66
Z_{wp1}	[mm]			20.4	20.6	16.8	17.0	14.9	13.9
t_{wp2}	[s]					1.42		1.18	1.10
Z_{wp2}	[mm]					10.5		7.1	6.1
t_{wp3}	[s]					1.89		1.75	1.61
Z_{wp3}	[mm]					15.9		12.8	12.2
t_{wp4}	[s]					2.47		2.14	2.06
Z_{wp4}	[mm]					10.1		7.7	6.4
t_{wp5}	[s]					2.47		2.14	2.06
Z_{wp5}	[mm]					13.0		11.9	11.3
t_{wp6}	[s]					3.57		3.26	3.12
Z_{wp6}	[mm]					9.8		9.4	8.6
t_{wp7}	[s]					4.12		3.92	3.64
Z_{wp7}	[mm]					12.2		11.1	10.7
t_{wp8}	[s]					4.62			4.09
Z_{wp8}	[mm]					10.8			10.0
u_{wr}	[mm/s]			21.94	23.68	21.82	26.98	23.28	21.06
u_{wp12}	[mm/s]					-9.8		-14.5	-17.6
u_{wp23}	[mm/s]					11.4		10.0	11.8
u_{wp34}	[mm/s]					-10.1		-13.0	-12.9
u_{wp45}	[mm/s]					5.1		6.5	7.4
u_{wp56}	[mm/s]					-5.8		-5.5	-8.6
u_{wp67}	[mm/s]					4.2		2.6	4.1
u_{wp78}	[mm/s]					-2.8			-1.6
u_{wr}	[mm/s]			21.94	23.68	21.82	26.98	23.28	21.06
$|u_{wom}|$	[mm/s]					7.0			9.1
$u_{wom,rec}$	[mm/s]					-7.1			-10.2
$|u_{wom,1-7}|$	[mm/s]					7.8		8.7	10.4
$u_{wom,1-6,rec}$	[mm/s]					-8.6		-11.0	-13.1

Table A.19. Values of the reorientation characteristics of the interface with liquid neon, hydrogen and cylinder B.

$\Delta T_w/\Delta Z$ [K/mm]		0.0	0.2	0.8	0.0	0.3	0.7
Exp.	ID	N1	N2	N3	H2	H1	H3
t_{cp1}	[s]	1.43	1.37	1.36	0.59	0.53	0.53
Z_{cp1}	[mm]						
t_{cp2}	[s]	4.45	4.03	3.62	1.62	1.56	1.47
Z_{cp2}	[mm]						
t_{cp3}	[s]				2.27	2.10	2.14
Z_{cp3}	[mm]						
t_{cp4}	[s]				3.32	2.94	3.01
Z_{cp4}	[mm]						
t_{cp5}	[s]				4.07	3.65	3.63
Z_{cp5}	[mm]						
t_{cp6}	[s]					4.42	4.55
Z_{cp6}	[mm]						
ω_d	[1/s]	1.04	1.18	1.39	3.77	4.22	4.02

A.2.2 Tables with the values of the pressure and temperature characteristics

Table A.20. Pressure and temperature values at the end of the experiments with liquid argon with cylinder A.

$\Delta T_w/\Delta Z$	P_v	T_S	T_{wf}	T_l	T_{wl}	T_v	T_{wv}	Experiment
[K/mm]	[hPa]	[K]	[K]	[K]	[K]	[K]	[K]	
0.15	966.4	86.9	88.0	86.7	87.4	96.4	98.8	A1
0.73	1110.2	88.2	90.0	84.8	132.5	86.6	141.3	A10
1.34	1532.4	91.4	97.8	85.0	86.5	177.0	192.1	A5

Table A.21. Pressure and temperature values at the end of the experiments with liquid argon with cylinder B. No estimation of the liquid temperature could be made.

$\Delta T_w/\Delta Z$ [K/mm]	P_v [hPa]	T_S [K]	T_L [K]	T_{WIF} [K]	T_{WL1} [K]	T_{WL2} [K]	T_{WV1} [K]	T_{WV2} [K]	T_{WV3} [K]	T_{WV4} [K]	T_{WV5} [K]	T_{V1} [K]	T_{V2} [K]	T_{V3} [K]	T_{V4} [K]	Experiment
1.4	1385.0	90.4	-	91.7	86.3	88.4	99.6	108.1	121.3	130.5	139.4	116.1	117.4	119.1	119.2	A14
2.0	1274.5	89.5	-	93.5	84.9	87.8	104.2	119.4	135.2	149.1	163.7	129.0	132.9	134.7	132.8	A16
2.5	1448.9	90.8	-	94.8	85.5	88.3	109.2	129.0	147.4	164.4	182.2	135.3	140.0	144.6	141.6	A15

Table A.22. Pressure and temperature values at the end of the experiments with liquid methane with cylinder B. No estimation of the liquid temperature could be made.

$\Delta T_w/\Delta Z$ [K/mm]	P_v [hPa]	T_S [K]	T_L [K]	T_{WIF} [K]	T_{WL1} [K]	T_{WL2} [K]	T_{WV1} [K]	T_{WV2} [K]	T_{WV3} [K]	T_{WV4} [K]	T_{WV5} [K]	T_{V1} [K]	T_{V2} [K]	T_{V3} [K]	T_{V4} [K]	Experiment
0.2	510.1	103.9		103.3	102.4	102.9	104.4	105.3	106.7	108.1	110.3	107.9	108.2	108.9	108.7	M24†
0.7	107.8	107.8		107.0	103.5	105.4	110.7	114.4	118.0	126.1	130.9	119.8	121.0	122.2	122.0	M23†
1.3	973.0	111.2		110.6	102.9	107.3	118.2	125.2	136.8	147.9	156.4	132.5	136.1	138.8	138.8	M18
1.3	1144.9	113.2		112.6	105.0	109.4	119.9	126.7	137.5	148.2	156.6	132.7	136.7	139.4	139.3	M19
1.9	1223.7	114.0		114.1	105.1	109.6	125.2	136.8	154.8	167.8	180.0	145.0	151.8	157.7	154.8	M17
1.9	1026.5	111.8		111.7	103.5	107.4	122.6	133.9	152.3	166.1	180.3	144.2	150.4	155.4	153.2	M20
2.5	990.1	111.4		112.3	100.3	105.9	127.2	144.7	166.8	183.4	200.4	156.5	162.2	166.9	166.6	M21
2.9	937.9	110.7		112.1	99.4	104.9	129.2	151.1	175.0	194.6	215.2	161.4	170.3	178.9	174.0	M22

Table A.23. Pressure and temperature values at the end of the experiments with liquid neon with cylinder B.

$\Delta T_w/\Delta Z$ [K/mm]	P_v [hPa]	T_S [K]	T_{WIF} [K]	T_{WL1} [K]	T_{WL2} [K]	T_{WV1} [K]	T_{WV2} [K]	T_{WV3} [K]	T_{WV4} [K]	T_{WV5} [K]	T_{V1} [K]	T_{V2} [K]	T_{V3} [K]	T_{V4} [K]	Experiment
0.0	1297	27.9	28.3	27.0	28.2	28.3	28.3	28.6	28.9	28.9	29.5	29.6	30.2	31.0	N1
0.2	965	26.9	27.5	27.0	27.3	28.2	29.0	30.5	32.5	33.8	33.0	33.5	34.6	35.2	N2
0.8	1144	27.5	28.2	26.8	27.3	31.1	34.9	39.9	47.3	53.6	39.8	43.9	48.3	50.6	N3

Table A.24. Pressure and temperature values at the end of the experiments with liquid hydrogen with cylinder B.

$\Delta T_w/\Delta Z$ [K/mm]	P_v [hPa]	T_S [K]	T_{WIF} [K]	T_{WL1} [K]	T_{WL2} [K]	T_{WV1} [K]	T_{WV2} [K]	T_{WV3} [K]	T_{WV4} [K]	T_{WV5} [K]	T_{V1} [K]	T_{V2} [K]	T_{V3} [K]	T_{V4} [K]	Experiment
0.0	1090	20.5	20.5	18.2	20.3	20.6	20.6	20.6	20.6	20.6	21.1	21.1	21.4	22.4	H2
0.3	782	19.4	19.8	18.9	19.6	20.2	20.5	21.0	22.4	22.1	25.5	25.3	25.6	30.9	H1
0.7	939	20.0	20.0	18.7	19.5	21.4	23.1	25.7	32.1	32.7	25.9	29.4	33.0	43.1	H3

A.2.3 Plots of the temperature evolutions

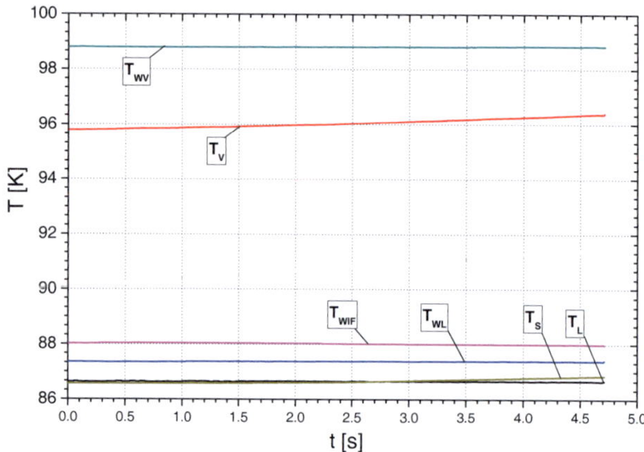

Fig. A.1. Evolutions of the liquid, vapor, wall and saturation temperatures for the experiment A1 with liquid argon.

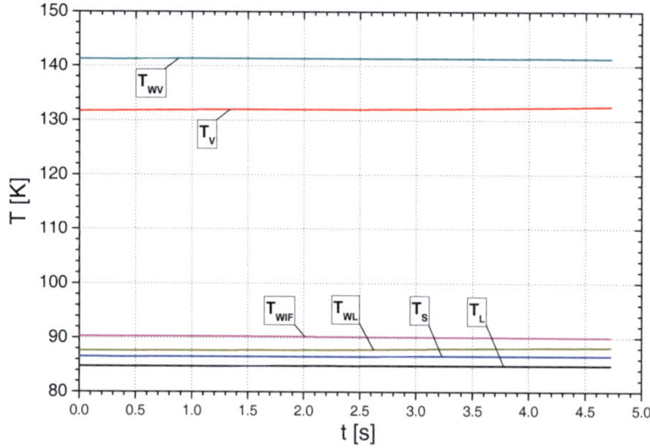

Fig. A.2. Evolutions of the liquid, vapor, wall and saturation temperatures for the experiment A10 with liquid argon.

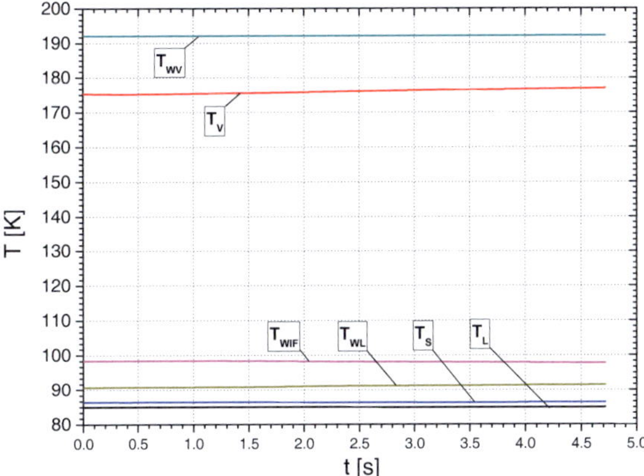

Fig. A.3. Evolutions of the liquid, vapor, wall and saturation temperatures for the experiment A5 with liquid argon.

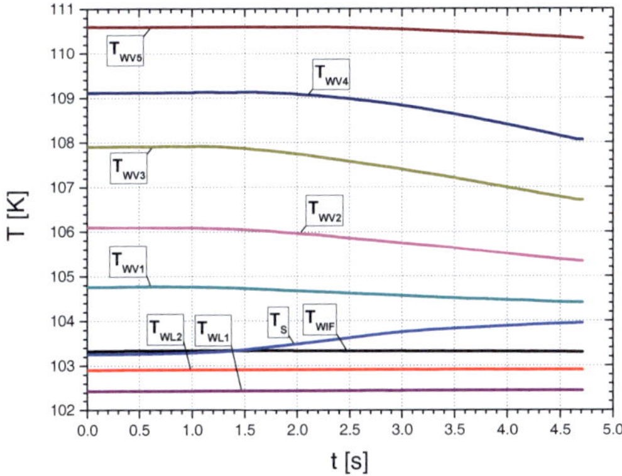

Fig. A.4. Evolutions of the wall and saturation temperatures for the experiment M24 with liquid methane.

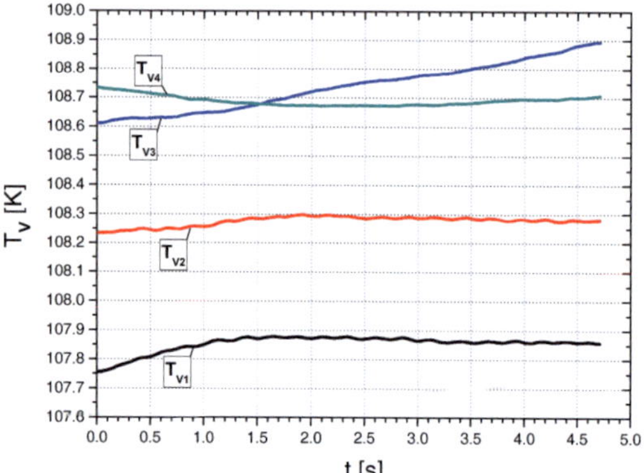

Fig. A.5. Evolution of the vapor temperature for the experiment M24 with liquid methane.

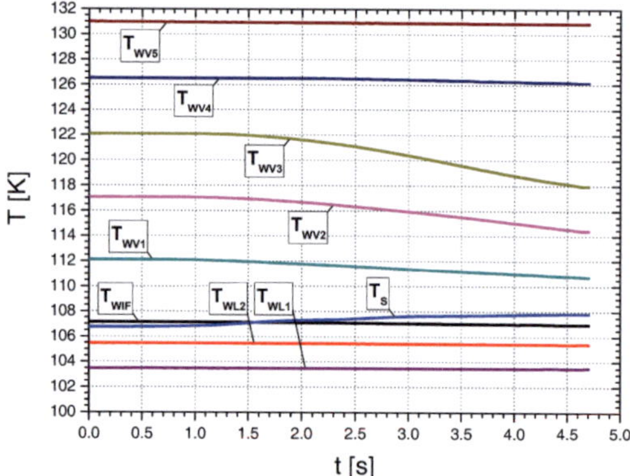

Fig. A.6. Evolutions of the wall and saturation temperatures for the experiment M23 with liquid methane.

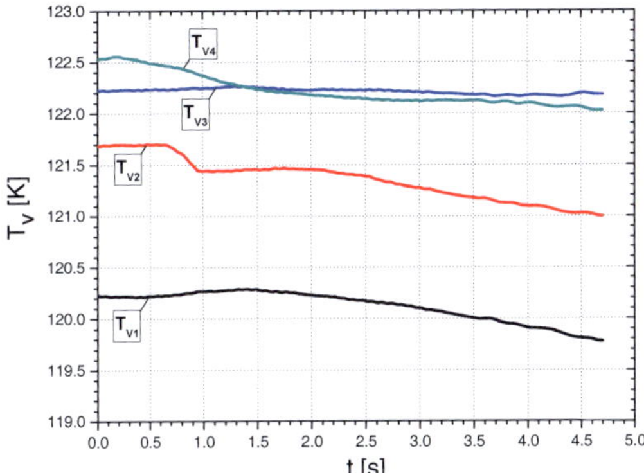

Fig. A.7. Evolution of the vapor temperature for the experiment M23 with LCH$_4$.

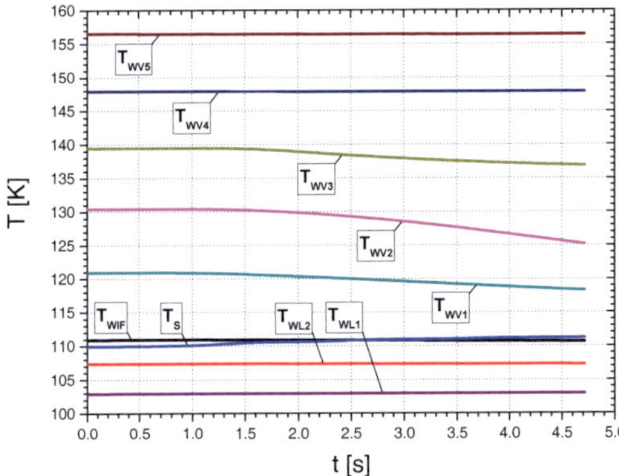

Fig. A.8. Evolutions of the wall and saturation temperatures for the experiment M18 with liquid methane.

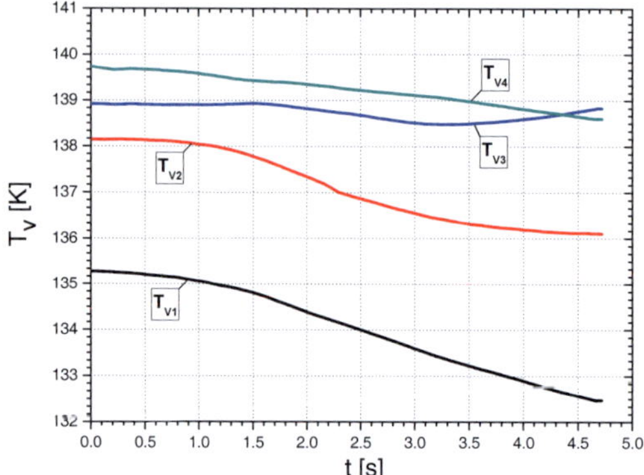

Fig. A.9. Evolution of the vapor temperature for the experiment M18 with liquid methane.

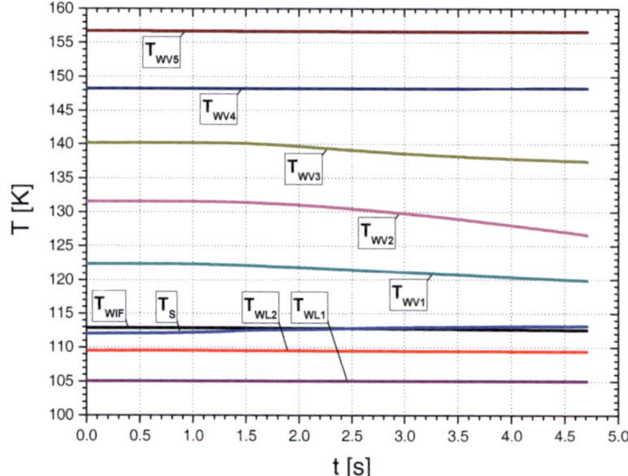

Fig. A.10. Evolutions of the wall and saturation temperatures for the experiment M19 with liquid methane.

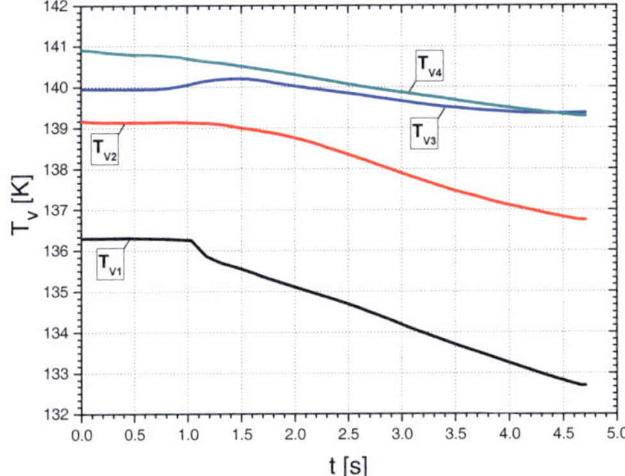

Fig. A.11. Evolution of the vapor temperature the by the experiment M19 with liquid methane.

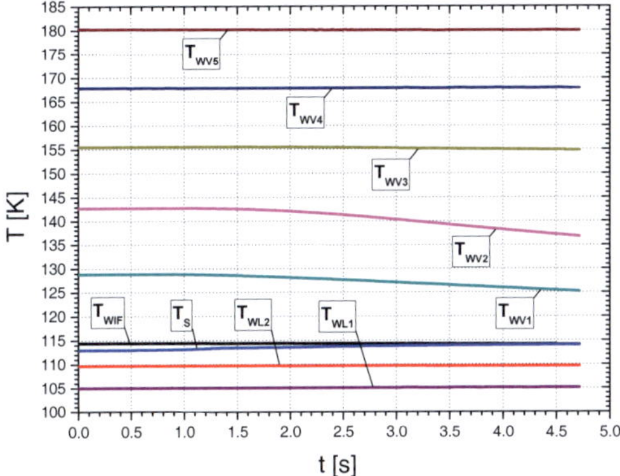

Fig. A.12. Evolutions of the wall and saturation temperatures for the experiment M17 with liquid methane.

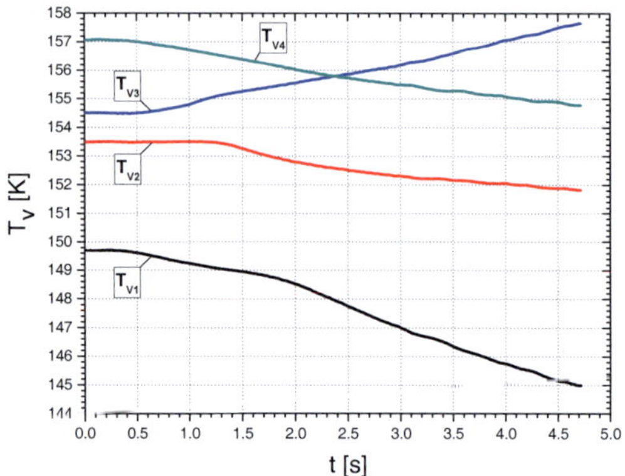

Fig. A.13. Evolution of the vapor temperature for the experiment M17 with liquid methane.

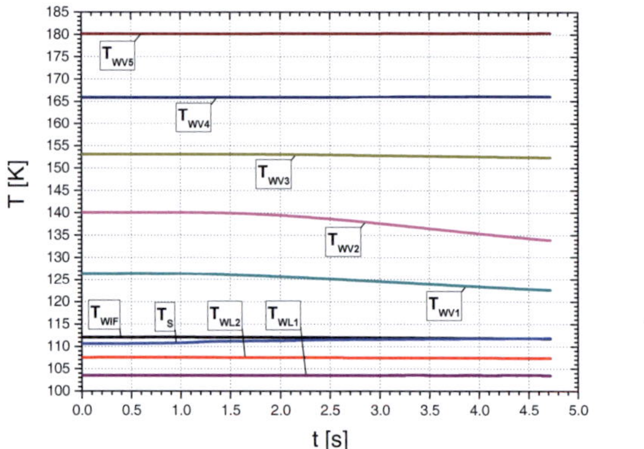

Fig. A.14. Evolutions of the wall and saturation temperatures for the experiment M20 with liquid methane.

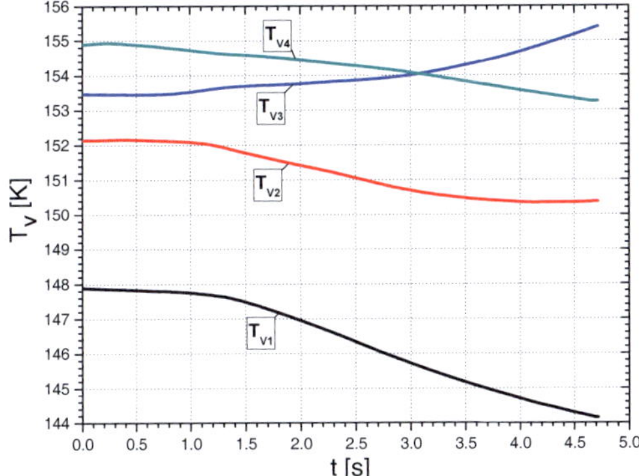

Fig. A.15. Evolution of the vapor temperature for the experiment M20 with liquid methane.

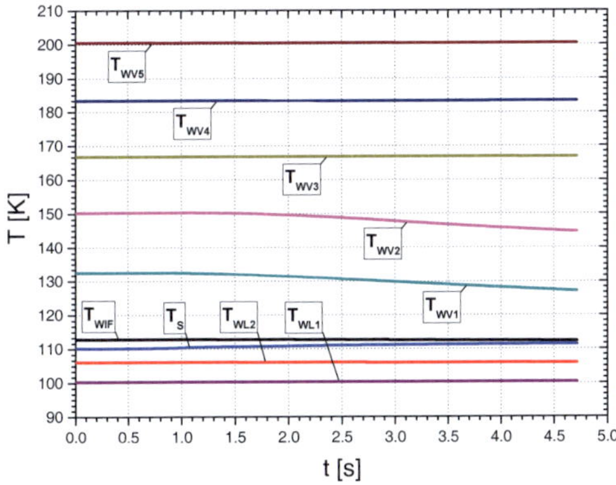

Fig. A.16. Evolutions of the wall and saturation temperatures for the experiment M21 with liquid methane.

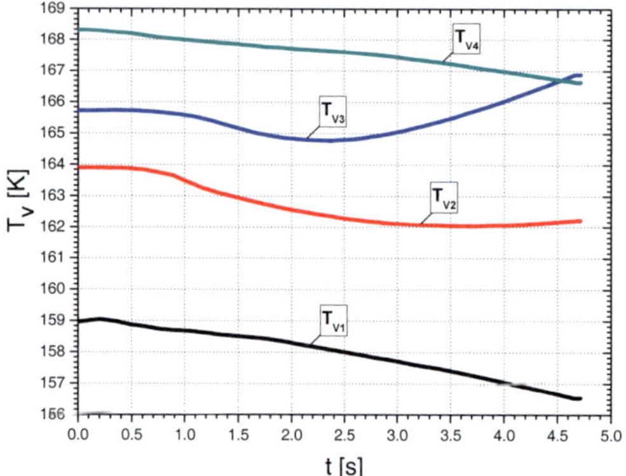

Fig. A.17. Evolution of the vapor temperature for the experiment M21 with liquid methane.

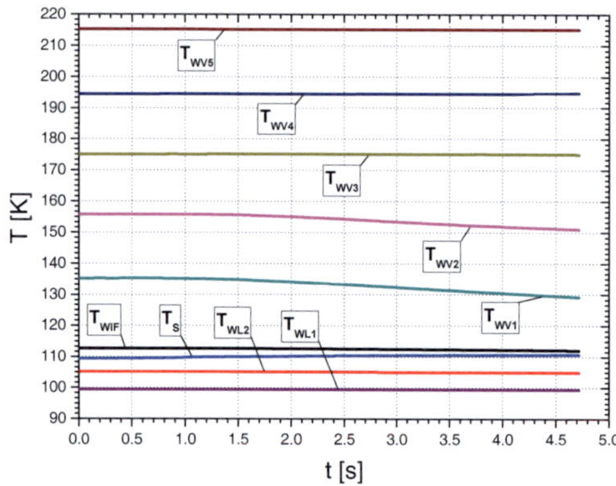

Fig. A.18. Evolutions of the wall and saturation temperatures for the experiment M22 with liquid methane.

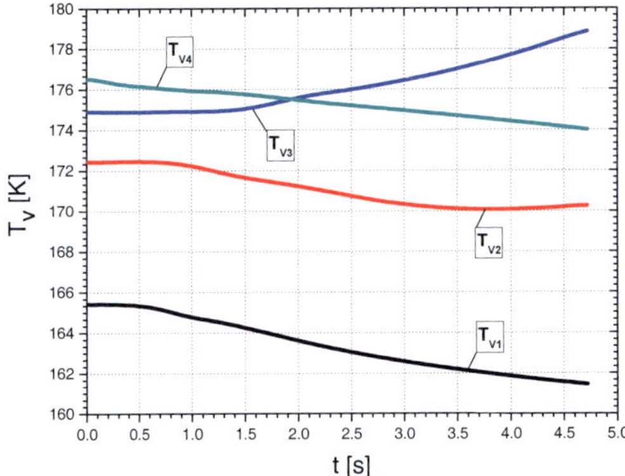

Fig. A.19. Evolution of the vapor temperature for the experiment M22 with liquid methane.

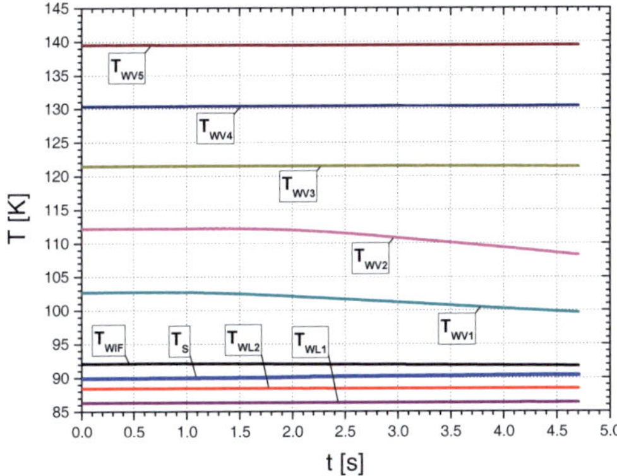

Fig. A.20. Evolutions of the wall and saturation temperatures for the experiment A14 with liquid argon.

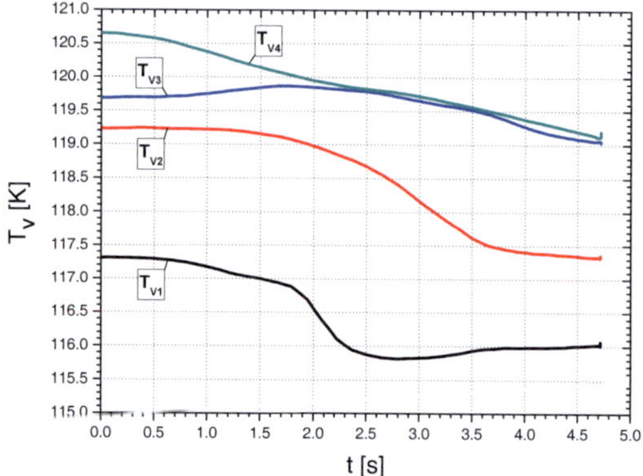

Fig. A.21. Evolution of the vapor temperature for the experiment A14 with liquid argon.

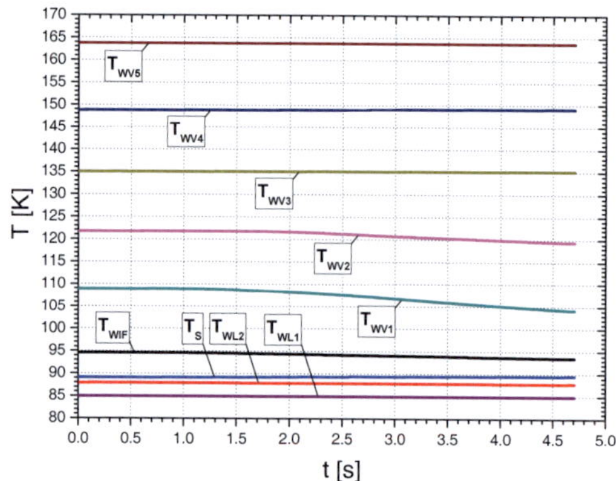

Fig. A.22. Evolutions of the wall and saturation temperatures for the experiment A16 with liquid argon.

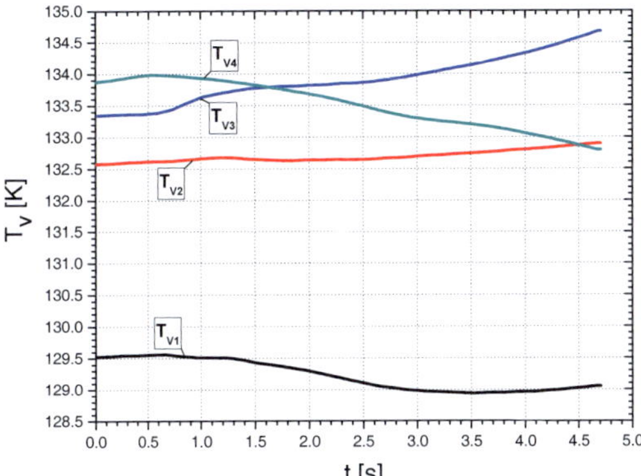

Fig. A.23. Evolution of the vapor temperature for the experiment A16 with liquid argon.

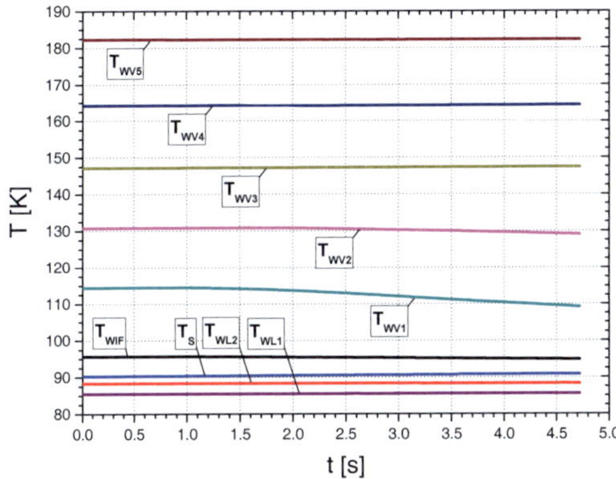

Fig. A.24. Evolutions of the wall and saturation temperatures for the experiment A15 with liquid argon.

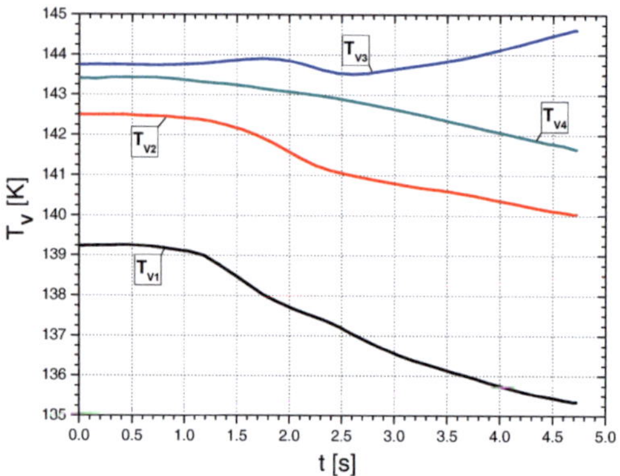

Fig. A.25. Evolution of the vapor temperature for the experiment A15 with liquid argon.

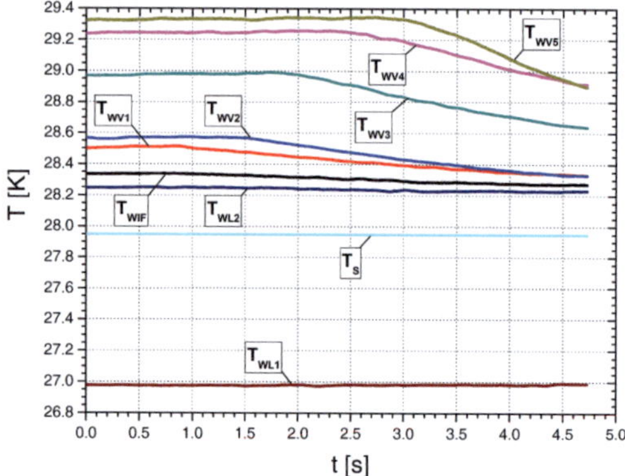

Fig. A.26. Evolutions of the wall and saturation temperatures for the experiment N1 with liquid neon.

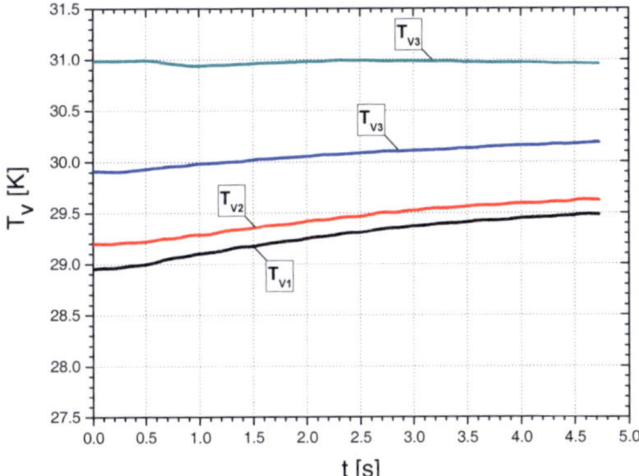

Fig. A.27. Evolution of the vapor temperature for the experiment N1 with liquid neon.

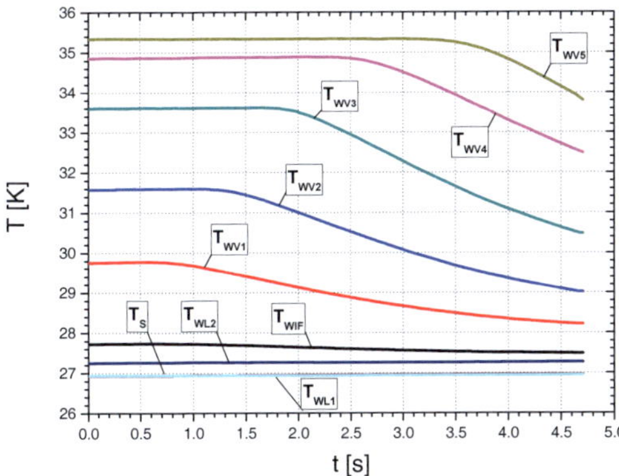

Fig. A.28. Evolutions of the wall and saturation temperatures for the experiment N2 with liquid neon.

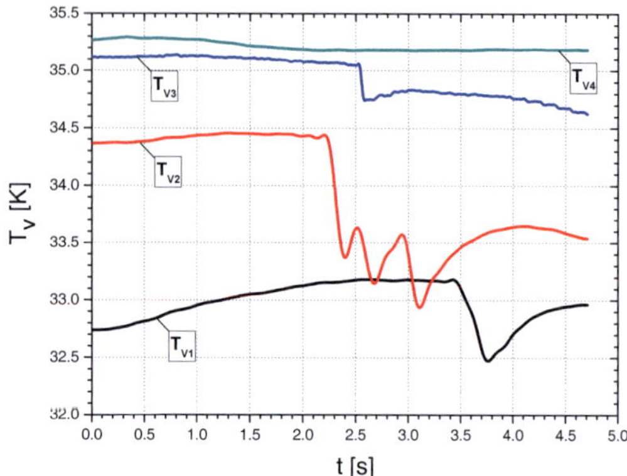

Fig. A.29. Evolution of the vapor temperature for the experiment N2 with liquid neon.

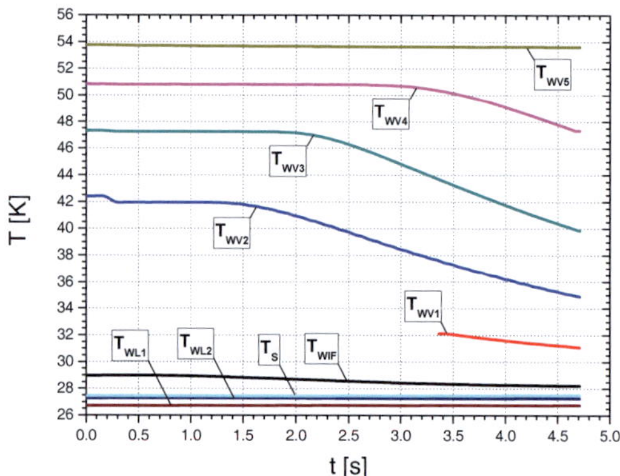

Fig. A.30. Evolutions of the wall and saturation temperatures for the experiment N3 with liquid neon.

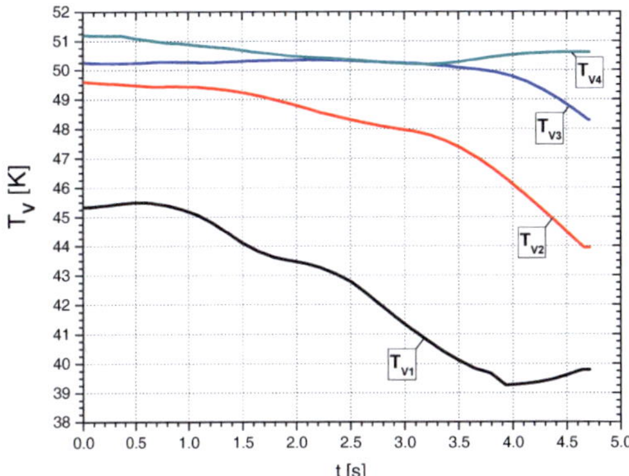

Fig. A.31. Evolution of the vapor temperature for the experiment N3 with liquid neon.

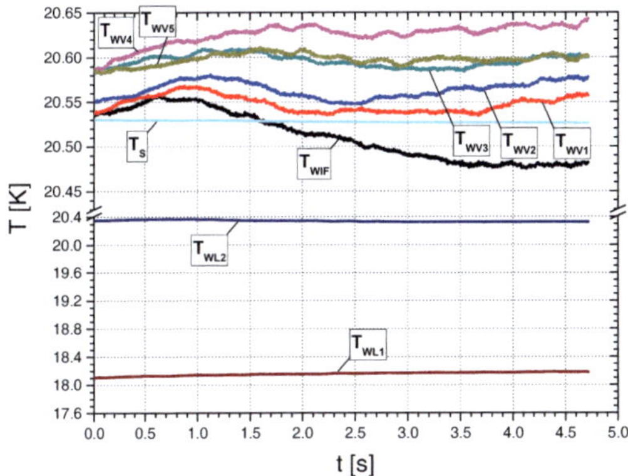

Fig. A.32. Evolutions of the wall and saturation temperatures for the experiment H2 with liquid hydrogen.

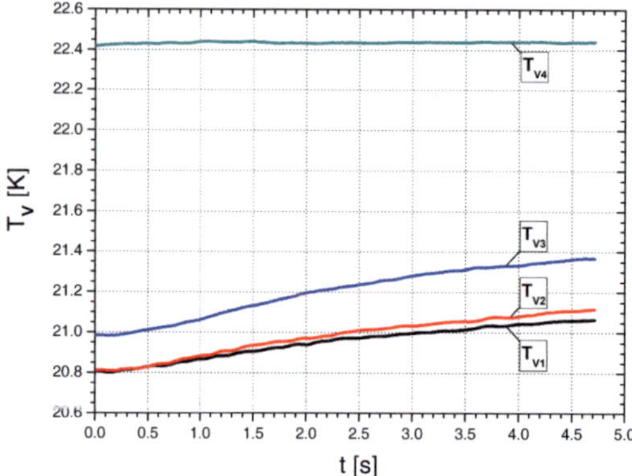

Fig. A.33. Evolution of the vapor temperature for the experiment H2 with liquid hydrogen.

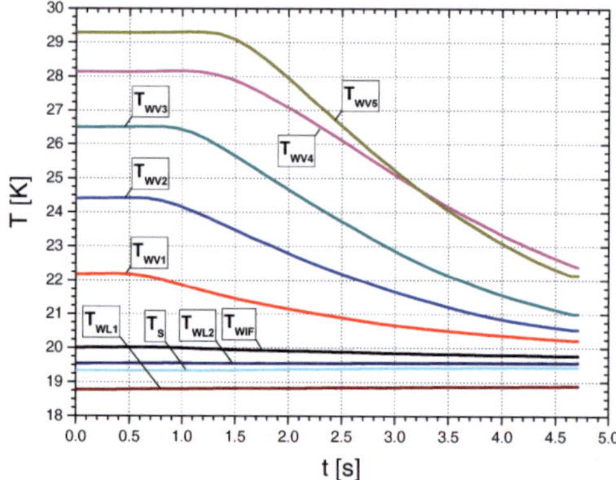

Fig. A.34. Evolutions of the wall and saturation temperatures for the experiment H1 with liquid hydrogen.

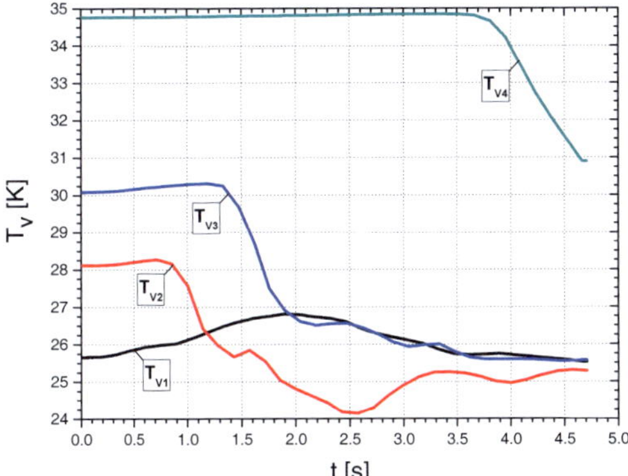

Fig. A.35. Evolution of the vapor temperature for the experiment H1 with liquid hydrogen.

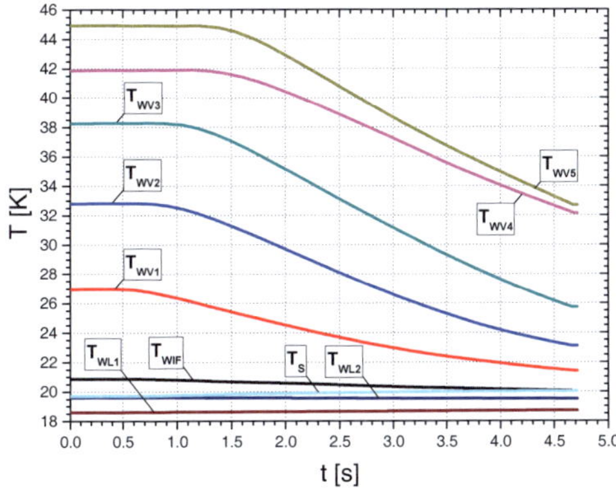

Fig. A.36. Evolutions of the wall and saturation temperatures for the experiment H3 with liquid hydrogen.

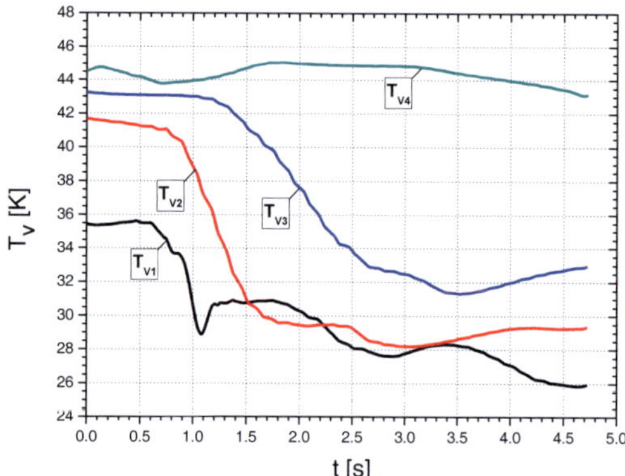

Fig. A.37. Evolution of the vapor temperature for the experiment H3 with liquid hydrogen.

A.3 Material properties

Table A.25. Material properties at 87.3 K with cylinder A.

Material	ρ [kg/m^3]	c_p [J/K-kg]	λ [W/m-K]	μ [Pa-s]	σ [N/m]
fused silica	2200.0	230.0	6.35×10^{-1}	-	-
LAr	1395.4	1117.2	1.30×10^{-1}	2.6×10^{-4}	1.25×10^{-2}
GAr	5.8	565.8	5.70×10^{-3}	7.0×10^{-6}	-

Table A.26. Material properties and their temperature dependence concerning the experiments with cylinder B. The data for the fluid phases is calculated with the NIST database ([49]) at 1013 hPa. The data for the solid phases is calculated through the CryoComp software ([22]). The properties of the borosilicate glass are presented through the data for Pyrex glass.

Material	T [K]	ρ [kg/m^3]	μ [Pa-s]	σ [N/m]	λ [W/m-K]	c_p [J/K-kg]	Δh_{lv} [J/kg]	ζ [-]
LAr	84	1415.7	2.9×10^{-4}		1.35×10^{-1}	1117.2		2.03
LAr	(sat) 87.3	1395.4	2.6×10^{-4}	1.25×10^{-2}	1.30×10^{-1}	1117.2	1.61×10^5	2.08
GAr	(sat) 87.3	5.8	7.0×10^{-6}		5.70×10^{-3}	565.8		1.72
GAr	107	4.6	8.7×10^{-6}		6.71×10^{-3}	541.7		1.70
GAr	127	3.9	10.1×10^{-6}		7.93×10^{-3}	532.5		1.69
GAr	145	3.4	11.6×10^{-6}		9.07×10^{-3}	528.5		1.68
LCH$_4$	96	444.4	1.7×10^{-4}		2.05×10^{-1}	3388.3		1.59
LCH$_4$	(sat) 111.7	422.4	1.2×10^{-4}	1.33×10^{-2}	1.84×10^{-1}	3481.1	5.11×10^5	1.69
GCH$_4$	(sat) 111.7	1.8	4.5×10^{-6}		1.20×10^{-2}	2217.7		1.37
GCH$_4$	134	1.5	5.4×10^{-6}		1.41×10^{-2}	2139.0		1.36
GCH$_4$	157	1.3	6.2×10^{-6}		1.69×10^{-2}	2112.7		1.35
GCH$_4$	180	1.1	7.1×10^{-6}		1.96×10^{-2}	2104.4		1.34
LNe	25	1244.2	1.46×10^{-4}		1.68×10^{-1}	2004.0		2.19
LNe	(sat) 27.1	1207.0	1.2×10^{-4}	0.45×10^{-2}	1.55×10^{-1}	1862.1	8.58×10^5	2.09
GNe	(sat) 27.1	9.6	4.5×10^{-6}		7.88×10^{-2}	1415.2		1.6
GNe	36	7.0	6.0×10^{-6}		0.90×10^{-2}	1074.0		1.71
GNe	44	5.7	7.3×10^{-6}		1.12×10^{-2}	1054.1		1.69
GNe	52	4.8	8.5×10^{-6}		1.31×10^{-2}	1045.4		1.69
LH$_2$	15	76.1	2.2×10^{-5}		8.42×10^{-2}	7104.7		1.38
LH$_2$	(sat) 20.3	70.8	1.3×10^{-5}	0.19×10^{-2}	1.55×10^{-1}	9658.3	4.45×10^5	1.72
GH$_2$	(sat) 20.3	1.34	1.1×10^{-6}		1.70×10^{-2}	12232.4		1.86
GH$_2$	29	0.88	1.5×10^{-6}		2.39×10^{-2}	10868.4		1.74
GH$_2$	37	0.72	1.90×10^{-6}		2.98×10^{-2}	10629.0		1.71
GH$_2$	45	0.55	2.26×10^{-6}		3.52×10^{-2}	10560.7		1.69
Pyrex	20	2214.0			1.46×10^{-1}	27.4		
Pyrex	50	2214.0			2.80×10^{-1}	136		
Pyrex	80	2214.0			4.43×10^{-1}	237.0		
Pyrex	130	2214.0			6.79×10^{-1}	357.0		
Pyrex	180	2214.0			8.50×10^{-1}	502.0		
Pyrex	220	2214.0			9.39×10^{-1}	570.0		
steel	100	7900.0			9.4	250.0		
steel	150	7900.0			11.5	347.0		
steel	200	7900.0			13.0	419.0		

Table A.27. Material properties of the fluid phases at the saturation temperature T_S corresponding to the vapor pressure P_v prior to the reduction of gravity with liquid argon with geometry A.

$\Delta T_w/\Delta Z$	ρ	μ	σ	λ	c_p	Δh_{lv}	λ_v	Experiment
[K/mm]	[kg/m^3]	[10^{-4} Pa-s]	[10^{-2} N/m]	[10^{-1} W/m-K]	[J/K-kg]	[10^5 J/kg]	[10^{-3} W/m-K]	
0.15	1400.3	2.67	1.27	1.31	1116.5	1.62	5.67	A1
0.73	1392.9	2.57	1.24	1.30	1117.7	1.61	5.73	A10
1.34	1373.6	2.35	1.17	1.25	1122.8	1.58	5.91	A5

Table A.28. Material properties of the fluid phases at the saturation temperature T_S corresponding to the vapor pressure P_v prior to the reduction of gravity with liquid argon with geometry B.

$\Delta T_w/\Delta Z$	ρ	μ	σ	λ	c_p	Δh_{lv}	λ_v	Experiment
[K/mm]	[kg/m^3]	[10^{-4} Pa-s]	[10^{-2} N/m]	[10^{-1} W/m-K]	[J/K-kg]	[10^5 J/kg]	[10^{-3} W/m-K]	
1.4	1379.3	2.41	1.19	1.26	1121.0	1.59	5.86	A14
2.0	1384.9	2.47	1.21	1.28	1119.5	1.60	5.80	A16
2.5	1376.7	2.38	1.18	1.26	1121.7	1.59	5.88	A15

Table A.29. Material properties of the fluid phases at the saturation temperature T_S corresponding to the vapor pressure P_v prior to the reduction of gravity with liquid methane with geometry B.

$\Delta T_w/\Delta Z$	ρ	μ	σ	λ	c_p	Δh_{lv}	λ_v	Experiment
[K/mm]	[kg/m^3]	[10^{-4} Pa-s]	[10^{-2} N/m]	[10^{-1} W/m-K]	[J/K-kg]	[10^5 J/kg]	[10^{-2} W/m-K]	
0.2	434.3	1.43	1.54	1.95	3426.7	5.25	1.04	M24
0.7	429.5	1.31	1.45	1.91	3447.3	5.20	1.09	M23
1.3	424.9	1.22	1.37	1.86	3468.5	5.14	1.13	M18
1.3	421.9	1.16	1.32	1.83	3483.5	5.10	1.16	M19
1.9	420.6	1.14	1.30	1.82	3490.2	5.09	1.18	M17
1.9	423.9	1.20	1.36	1.85	3473.4	5.13	1.14	M20
2.5	424.6	1.21	1.37	1.86	3469.9	5.14	1.14	M21
2.9	425.6	1.23	1.39	1.87	3465.1	5.15	1.13	M22

Table A.30. Material properties of the fluid phases at the saturation temperature T_S corresponding to the vapor pressure P_v prior to the reduction of gravity with liquid neon with geometry B.

$\Delta T_w/\Delta Z$	ρ	μ	σ	λ	c_p	Δh_{lv}	λ_v	Experiment
[K/mm]	[kg/m^3]	[10^{-4} Pa-s]	[10^{-3} N/m]	[10^{-1} W/m-K]	[J/K-kg]	[10^4 J/kg]	[10^{-3} W/m-K]	
0.0	1193.0	1.07	4.19	1.50	1888.5	8.48	8.66	N1
0.2	1210.5	1.19	4.61	1.56	1859.1	8.60	11.34	N2
0.8	1201.8	1.27	4.40	1.53	1869.6	8.54	9.53	N3

Table A.31. Material properties of the fluid phases at the saturation temperature T_S corresponding to the vapor pressure P_v prior to the reduction of gravity with liquid hydrogen with geometry B.

$\Delta T_w/\Delta Z$	ρ	μ	σ	λ	c_p	Δh_{lv}	λ_v	Experiment
[K/mm]	[kg/m^3]	[10^{-5} Pa-s]	[10^{-3} N/m]	[10^{-1} W/m-K]	[J/K-kg]	[10^5 J/kg]	[10^{-2} W/m-K]	
0.0	70.5	1.31	1.93	1.04	9815.3	4.44	1.73	A1
0.3	71.9	1.44	2.12	1.02	9036.9	4.49	1.60	A10
0.7	71.4	1.39	2.06	1.03	9290.4	4.48	1.64	A5